ANNALS OF THE NEW YORK ACADEMY OF SCIENCES

Volume 894

EDITORIAL STAFF

Executive Editor
BARBARA M. GOLDMAN

Managing Editor
JUSTINE CULLINAN

Associate Editor
MARY KATHERINE BRENNAN

The New York Academy of Sciences
2 East 63rd Street
New York, New York 10021

THE NEW YORK ACADEMY OF SCIENCES
(Founded in 1817)

BOARD OF GOVERNORS, September 15, 1999 – September 15, 2000

BILL GREEN, *Chairman of the Board*
TORSTEN WIESEL, *Vice Chairman of the Board*
RODNEY W. NICHOLS, *President and CEO* [ex officio]

Honorary Life Governors
WILLIAM T. GOLDEN   JOSHUA LEDERBERG

JOHN T. MORGAN, *Treasurer*

Governors

D. ALLAN BROMLEY	LAWRENCE B. BUTTENWIESER	PRAVEEN CHAUDHARI
JOHN H. GIBBONS	RONALD L. GRAHAM	HENRY M. GREENBERG
ROBERT G. LAHITA	MARTIN L. LEIBOWITZ	JACQUELINE LEO
WILLIAM J. McDONOUGH	KATHLEEN P. MULLINIX	JOHN F. NIBLACK
SANDRA PANEM	RICHARD RAVITCH	RICHARD A. RIFKIND

SARA LEE SCHUPF   JAMES H. SIMONS

ELEANOR BAUM, *Past Chairman of the Board*
HELENE L. KAPLAN, *Counsel* [ex officio]   PETER H. KOHN, *Secretary* [ex officio]

FOOD AND AGRICULTURAL SECURITY
GUARDING AGAINST NATURAL THREATS AND TERRORIST ATTACKS AFFECTING HEALTH, NATIONAL FOOD SUPPLIES, AND AGRICULTURAL ECONOMICS

ANNALS OF THE NEW YORK ACADEMY OF SCIENCES
Volume 894

FOOD AND AGRICULTURAL SECURITY
GUARDING AGAINST NATURAL THREATS AND TERRORIST ATTACKS AFFECTING HEALTH, NATIONAL FOOD SUPPLIES, AND AGRICULTURAL ECONOMICS

Edited by Thomas W. Frazier and Drew C. Richardson

The New York Academy of Sciences
New York, New York
1999

Copyright © 1999 by the New York Academy of Sciences. All rights reserved. Under the provisions of the United States Copyright Act of 1976, individual readers of the Annals are permitted to make fair use of the material in them for teaching or research. Permission is granted to quote from the Annals provided that the customary acknowledgment is made of the source. Material in the Annals may be republished only by permission of the Academy. Address inquiries to the Executive Editor at the New York Academy of Sciences.

Copying fees: *For each copy of an article made beyond the free copying permitted under Section 107 or 108 of the 1976 Copyright Act, a fee should be paid through the Copyright Clearance Center, Inc., 222 Rosewood Drive, Danvers, MA 01923. The fee for copying an article is $3.00 for nonacademic use; for use in the classroom, it is $0.07 per page.*

⊗The paper used in this publication meets the minimum requirements of the American National Standard for Information Sciences—Permanence of Paper for Printed Library Materials, ANSI Z39.48-1984.

Library of Congress Cataloging-in-Publication Data

Food and agricultural security : guarding against natural threats and terrorist attacks affecting health, national food supplies, and agricultural economics / edited by Thomas W. Frazier and Drew C. Richardson
    p. cm. — (Annals of the New York Academy of Sciences ; v. 894)
    This volume is the result of a conference entitled International Conference on Food and Agricultural Security, September 28–30, 1998.
    Includes bibliographical references (p.).
    ISBN 1-57331-230-4 (cloth : alk. paper). — ISBN 1-57331-231-2 (pbk : alk. paper)
    1. Food industry and trade—Defense measures—Congresses. 2. Food industry and trade—Defense measures—United States—Congresses. 3. Agriculture—Defense measures—Congresses. 4. Agriculture—Defense measures—United States—Congresses. I. Frazier, Thomas W. II. Richardson, Drew C. III. International Conference on Food and Agricultural Security d (1998) IV. Series.

UA929.95.F6 F66 1999
338.1'973—dc21   99-057144

GYAT / B-M
Printed in the United States of America
ISBN 1-57331-230-4 (cloth)
ISBN 1-57331-231-2 (paper)
ISSN 0077-8923

ANNALS OF THE NEW YORK ACADEMY OF SCIENCES
Volume 894

FOOD AND AGRICULTURAL SECURITY
GUARDING AGAINST NATURAL THREATS AND TERRORIST ATTACKS AFFECTING HEALTH, NATIONAL FOOD SUPPLIES, AND AGRICULTURAL ECONOMICS[a]

Editors
THOMAS W. FRAZIER AND DREW C. RICHARDSON

Conference Organizers
THOMAS W. FRAZIER AND DREW C. RICHARDSON

CONTENTS

Natural and Bioterrorist/Biocriminal Threats to Food and Agriculture. *By* THOMAS W. FRAZIER	1
Agriculture and Food Security. *By* FLOYD P. HORN AND ROGER G. BREEZE	9
The Soviet Union's Anti-Agricultural Biological Weapons. *By* KENNETH ALIBEK	18
The Threat Posed by the Global Emergence of Livestock, Food-borne, and Zoonotic Pathogens. *By* FREDERICK A. MURPHY	20
Contemporary Global Movement of Emerging Plant Diseases. *By* RANAJIT BANDYOPADHYAY AND RICHARD A. FREDERIKSEN	28
Biological Warfare Training: Infectious Disease Outbreak Differentiation Criteria. *By* DONALD L. NOAH, ANNETTE L. SOBEL, STEPHEN M. OSTROFF, AND JOHN A. KILDEW	37
The U.S. Department of Agriculture Food Safety and Inspection Service's Activities in Assuring Biosecurity and Public Health Protection. *By* BONNIE BUNTAIN AND GEORGE BICKERTON	44

[a]This volume is the result of a conference entitled **International Conference on Food and Agricultural Security** held by the U.S. Department of Agriculture, FMI Scientific Laboratory, the Department of Defense Veterinary Service Activity, the American Veterinary Medical Association, Louisiana State University, and the National Consortium for Genomic Resources Management and Services on September 28–30, 1998 in Washington, D.C.

Safeguarding Production Agriculture and Natural Ecosystems against Biological Terrorism: A U.S. Department of Agriculture Emergency Response Framework. *By* RON SEQUEIRA 48

Guarding against Natural Threats and Terrorist Attacks: An Industry Perspective. *By* D. STOLTE AND K.E. OLSON 68

The Role of National Animal Health Emergency Planning. *By* JOHN B. ADAMS 73

Industry Concerns and Partnerships to Address Emerging Issues. *By* BETH LAUTNER 76

International Economic Considerations Concerning Agricultural Diseases and Human Health Costs of Zoonotic Diseases. *By* ALFONSO TORRES . 80

The Cost of Disease Eradication: Smallpox and Bovine Tuberculosis. *By* ANN MARIE NELSON 83

Economic Considerations of Agricultural Diseases. *By* CORRIE BROWN 92

Regionalization's Potential in Mitigating Trade Losses Related to Livestock Disease Entry. *By* ANN HILLBERG SEITZINGER, KENNETH W. FORSYTHE, JR., AND MARY LISA MADELL 95

Foreign Animal Disease Agents as Weapons in Biological Warfare. *By* DAVID R. FRANZ 100

Tools and Methods for Protection of Targets and Infrastructures Associated with Food and Agriculture Industries. *By* DAVID L. HUXSOLL 105

Infecting Soft Targets: Biological Weapons and Fabian Forms of Indirect Grand Strategy. *By* ROBERT D. HICKSON 108

The First Step toward Building Tools and Methods for Protection. *By* JOSEPH D. DOUGLASS, JR. 118

Targeted Immune Design Using RNA Immunization. *By* ROBERT R. GARRITY 124

Sensitive and Rapid Identification of Biological Threat Agents. *By* J.A. HIGGINS, M.S. IBRAHIM, F.K. KNAUERT, G.V. LUDWIG, T.M. KIJEK, J.W. EZZELL, B.C. COURTNEY, AND E.A. HENCHAL 130

United States of America v. Ray Wallace Mettetal, Jr.: Preliminary Observations. *By* RAY B. FITZGERALD, JR. 149

Terrorism Overview. *By* PETER S. PROBST 154

The Changing Biological Warfare Threat: Anti-Crop and Anti-Animal Agents. *By* SHARON A. WATSON 159

Trends in American Agriculture: Their Implications for Biological Warfare against Crop and Animal Resources. *By* WALLACE A. DEEN 164

Agroterrorism: Agricultural Infrastructure Vulnerability. *By* JURGEN VON BREDOW, MICHAEL MYERS, DAVID WAGNER, JAMES J. VALDES, LARRY LOOMIS, AND KAVEH ZAMANI 168

The Need for a Coordinated Response to Food Terrorism: The Wisconsin Experience. *By* NICHOLAS J. NEHER 181

The Threat of Bioterrorism to U.S. Agriculture. *By* MICHAEL V. DUNN 184

Where Have All My Pumpkins Gone?: The Vulnerability of Insect Pollinators. *By* BARRY H. THOMPSON 189

The Role of Pesticides in Agricultural Crop Protection. *By* NANCY N. RAGSDALE 199

The Status and Role of Vaccines in the U.S. Food Animal Industry: Implications for Biological Terrorism. *By* PETER L. NARA 206

Exotic Diseases of Citrus: Threats for Introduction? *By* RONALD H. BRLANSKY 218

What Should the G8 Do about the Biological Warfare Threat To International Food Safety? *By* WENDY BARNABY 222

Roundtable Summary: A Domestic Legislative Agenda for Improving Food Safety and Safeguards from Terrorist Attacks on the U.S. Food Supplies and U.S. Agricultural Interests. *By* LONNIE KING 228

Index of Contributors 233

Assistance was received from:

Major Funders
- FEDERAL BUREAU OF INVESTIGATION LABORATORY
- U.S. DEPARTMENT OF AGRICULTURE—AGRICULTURE RESEARCH SERVICE

Supporters
- AMERICAN VETERINARY MEDICAL ASSOCIATION
- DEPARTMENT OF DEFENSE VETERINARY SERVICE ACTIVITY
- LOUISIANA STATE UNIVERSITY

The New York Academy of Sciences believes it has a responsibility to provide an open forum for discussion of scientific questions. The positions taken by the participants in the reported conferences are their own and not necessarily those of the Academy. The Academy has no intent to influence legislation by providing such forums.

Natural and Bioterrorist/Biocriminal Threats to Food and Agriculture

THOMAS W. FRAZIER[a]

GenCon, 294 Woodland Drive, Montross, Virginia 22520, USA

We should all appreciate that the general area of weapons of mass destruction (WMD) threats to America is a very sensitive topic at this time.[1–3] It is so sensitive, in fact, that some organizations with keen interests were unwilling to even be listed as a sponsor organization or to send a speaker to this conference. They feared that their participation might increase the possibility that their organization would become a target for some domestic or transnational biocriminal or terrorist act. Many large American corporations are alarmed about terrorist threats to their operations and physical assets but will not express their sensitivities and concerns in public forums. This basic ambivalence requires resolution before any constructive remedies are possible.

There are significant government sensitivities too, both domestic and foreign. Some "friendly" governments are reluctant to discuss WMD terrorist issues in public. Reluctance seems rather self-defeating because an effective defense program requires cooperation among organizations. We cannot counter a threat that cannot be described and discussed. Of course, we appreciate that hoaxes, extortion threats, and false claims can precipitate heavy financial losses to companies and governments. Nevertheless, fear cannot be allowed to dominate reason and block constructive preparation and defensive action. Otherwise, terrorists such as Osama bin Laden and dictators such as Sadaam Hussein of Iraq will be encouraged to extend their campaigns of terrorism and irregular warfare even further.

When government organizations deny the existence of a legitimate threat or conceal specific bioterrorist or biocriminal actions affecting their constituencies, the subsequent uncovering of this denial or concealment adds to distrust of government and general discomfort. If a target population's fear and distrust of government's capability for protection reach substantial levels, terrorists have achieved most of their goals. Terrorism is undertaken to force unwelcome policy changes upon a population through intimidation and other aversive means—not just to harm specific people. With sufficient intimidation and successful undermining of confidence in the existing government, people become more amenable to entertaining political or economic propositions or violations of human rights that they would not otherwise accept.[4]

The American people and American research and development communities need to be provided with enough information to understand the basic contemporary problems of and vulnerabilities to WMD terrorism today.[5] Otherwise, there will be insufficient support from the population and opinionmakers when it comes to obtain-

---

[a]Address correspondence to: Thomas W. Frazier, Ph.D., President, GenCon, 294 Woodland Drive, Montross, Virginia 22520, Telephone: (804)472-3029; Fax: (804)472-9483.
e-mail: gencon@3N.net

ing the needed public support for launching substantive government WMD defense programs to protect the population and American assets here and abroad. Food and agriculture supply and safety represent basic American strengths, but terrorist strategies seek to convert strengths into handicaps. Food and agricultural targets and associated infrastructures can be attacked easily with little chance that a biological attack will be recognized for what it is. Some analysts suspect, in fact, that experiments in attacking food and agriculture targets may have been in progress for some time by enemies of America and its allies. We certainly won't find out much about the validity of this suspicion unless we are prepared to investigate possible incidents routinely with the object of differentiating between naturally occurring problems and man-made attacks.

The most recent incidents involving food targets are a string of food-poisoning cases in Japan. The major case involved arsenic poisoning of a curry dish at a festival in Wakayama, where dozens fell ill and four people died. Other poisoning incidents during this time period involved poisoned tea and canned drinks. The police have made no arrests and it is unclear whether these incidents include copy-cat poisonings or whether there is a terrorist campaign in progress. The apparent random nature of these terrorist actions is said to have enhanced terror reactions in the Japanese general population.[6]

As for attacks on food and agriculture infrastructures, naturally and maliciously introduced pathogens may attack their targets indirectly. Attacks may be made against food and agriculture transportation systems, on water supplies, on farm workers, on food handlers, and on processing facilities. Many past incidents have involved plots to infect food at grocery stores. False claims and hoaxes can be introduced to challenge public confidence in food safety for particular commodities. Insect hosts for diseases affecting crop plants or livestock may be imported and released with the intent of creating an epidemic or influencing a nation's ability to export agricultural products abroad. American reliance on imported fresh fruits and vegetables has grown so much that safety of imported foods has become a major source of public health concern.[7–10]

Shortly before this conference, I received an e-mail from Tim Harris, the European Secretary of the Animal Transportation Association. In this communication he stated, "Based here in UK as European Secretary of the Animal Transportation Association I am acutely aware of attacks on all sorts of commercial biological targets from GM crops to livestock haulers, veterinary laboratories, breeding establishments, and veterinary surgeries, to name but a few."

To summarize, we are apparently experiencing contemporary global assaults upon humans, livestock, crops, and the environment itself. These assaults have either natural elements or deliberate malicious elements that may be difficult or impossible to differentiate from one another. We are working more or less in the dark because governments and companies use very compartmentalized security measures to keep one another and the public in the dark. However, due largely to the media and public-spirited persons, the legitimacy and scope of these assaults are coming into public awareness. Unnecessary official denial and evasion have to be rejected now and replaced by candid public debate and far better information sharing among governments and individual agencies. Some official agreement must be reached about what is necessary and desirable to communicate to the public and what is not. Conferences of the present sort can provide a useful means for preparing the population to understand prospective threats and help build support for informed and constructive action. The science and technology communities and the general public have much to offer and need to be able to hold officials accountable for their actions or inaction.

NEED FOR A BROAD CONCEPTUAL SCHEMA AND STRATEGIC PLAN FOR DEFENDING FROM NATURAL AND MAN-MADE THREATS

Whether damage or illness is naturally or intentionally introduced, the remedies may be identical, assuming that remedies are possible. Most of the pathogens used in biowarfare or bioterrorism are obtained simply through purchase of existing cultures of the pathogen. Of course, now a biological pathogen can be genetically modified to make it less amenable to treatment or for some other reason. But microorganisms mutate on their own—as human immunodeficiency virus (HIV) has amply demonstrated—through such a rapid evolution that the mutations seem to outpace the rate at which treatments can be developed. We should also remember that many serious communicable diseases are spread via animal hosts, insects, or tainted food or water containing dangerous poisons or parasites.

It makes sense to look at these biological threats in a spectral or behavioral gradient context—perhaps a gradient that classifies threats according to increasing levels of criminality or intentional damage to living organisms and their support system infrastructures. This gradient would start with legal, but unsustainable or damaging practices, such as depletion of rainforests for subsistence farming, which quickly destroys fragile tropical soils and leads to erosion and desertification. It creates scrub areas that operate as breeding grounds for disease-carrying insects. Then there are practices that are small offenses according to the law, but which are destructive nonetheless, such as overgrazing, overfishing, illegal water use, or disregard of food and drug safety regulations imposed by governments. These activities are punished because they are unsustainable or because they would inevitably create widespread, even irreversible, damage if unchecked.

At a more seriously criminal position on the gradient would be such activities as smuggling dangerous substances and illegal specimens and the production and distribution of dangerous drugs for purposes of personal gain. Food tampering and extortion threats involving poisoning water supplies, food, or drugs also lie in this mid-area of the gradient scale. Incidents such as assassinations and terrorist threats and activities would be listed as still more serious crimes. The most dangerous activities would be direct state employment of weapons of mass destruction against human, food, or agricultural targets as acts of genocide of ethnic populations or WMD attacks as acts of war against other countries.

If some such schema were to be accepted, then a number of significant considerations might become more apparent. In the U.S., these notably include the need for developing stronger links among Commerce, Justice, USDA, FDA, and DOD, and the larger law enforcement and intelligence communities for timely and useful information sharing, reallocation of regulatory responsibilities, and national security and law enforcement collaborations. Jurisdictional problems in regulation and enforcement might be more easily overcome. This emphasis on cooperative linkages should be an important part of our defense-of-the-homeland agenda. Fortunately there is good initial progress in the U.S. to enhance these cooperative linkages and to create important functional linkages with state and local governments. These linkages must be extended internationally in view of the natural international spread of diseases, transnational terrorism, and international biocrimes.

The biggest problem for developing a strategic view of the biological threats problem is the small number of specialists in the U.S. who fully understand what terrorism involves and what options biological warfare tools can make available to terrorists.[11]

The U.S. program as a whole, however, remains largely reactive, compartmentalized, and susceptible to impulsive congressional reactions when a terrorist event occurs.[11] What is needed is a planning platform based upon what kinds of terrorists and terrorism can be expected in the future. These issues have been well studied and presented to the U.S. Congress. The Nunn-Lugar-Domenici legislation was a significant milestone in addressing the WMD terrorist problem. However, those few senators who have shown some leadership keep having their attention diverted to other issues. The training of first responders in crisis and consequences management is hobbled by the lack of significant information about potential threats. Information about the nature of WMD has been classified and can't be divulged to them. Other potential first responders to bioterrorism receive no training at all in crisis and consequences management for a large biowarfare incident or even in recognizing the symptoms associated with particular prospective agents.

Therefore, there is a clear need to develop effective educational programs for stimulating continued attention of congressional decisionmakers, for alerting companies that may be perceived as infrastructure targets to terrorists, and for training first responders who will come into contact with affected people, pets, or livestock after an bioterrorism attack occurs.

A number of outstanding committee staff in the Senate and House of Representatives can be contacted on technical and legislative issues. Richard Clarke is the National Security Council Director for Global Issues and Multilateral Affairs and chairs counterterrorism coordination activities. The Senate Permanent Committee on Investigations has played an especially important leadership role and continues to have hearings on safety and security issues. Senator Jon Kyl's subcommittee on Technology, Terrorism and Government Information is rapidly becoming an asset in the Senate. Senator Lugar has shown good past leadership and may become re-engaged in the WMD terrorism defense area. The food safety area has also prompted some especially significant legislative initiatives.

Senator Lugar has brought attention to "reverse bioterrorism" in his most recent initiative introducing an amendment (No. 3145) to the Agriculture Appropriations Act of 1999.[12] The purpose of this amendment was "To Provide a Framework for Consideration by the Legislative and Executive Branches of Unilateral Economic Sanctions in Order To Ensure Coordination of United States Policy with Respect to Trade Security, and Human Rights." The amendment dealt with what some call reverse bioterrorism, or the employment of unilateral sanctions to prohibit the provision of food or medicines to specified nations and other sanction measures. Senator Lugar pointed to a dramatic rise in the number and variety of U.S. economic sanctions directed against other nations. These unilateral sanctions are often counterproductive to our interests and may cause unintended effects.

The amendment was tabled 53–46 and was followed by an amendment by Senator Dodd that passed. However, Senator Torricelli added a second-degree amendment that essentially blocked most of the specific suggestions originally presented by Senator Lugar.[12]

So while the U.S. no longer employs chemical and biological weapons against other countries, it still does employ unilateral sanctions very liberally, which can achieve similar effects in some cases where poverty and disease are prevalent. It is the possibility of economically oriented terrorism on the part of major WMD powers that presents the most disturbing prospects for the future. Next-century economic warfare involving introduction of unilateral economic sanctions, combined with selective introduction of pathogens to destroy already fragile economic sectors of target governments or to manipulate foreign ethnic populations in the quest for cheap earth resources, could give an entirely new face to colonial conquest. This could, of course, set the stage for retaliation by third-world-associated terrorist groups using chemical and biological weapons easily bought, transported, and released onto targets with little danger of discovery or apprehension.

These speculations actually are not at all futuristic or hypothetical. The U.S. has been squeezing Cuba through economic sanctions for years. There are claims that in the late 1970s a U.S. intelligence agent provided a terrorist organization with a pathogen for introducing foot-and-mouth disease in Cuban swine. A serious epidemic robbed Cuba of an important staple in the Cuban diet. These claims have been denied, but there were plots at this time to assassinate Castro using biological agents.[13,14] Now, in 1998, Cuba has its own biowarfare and chemical warfare program integrated into very modern biotechnology facilities. These costly facilities are a great burden on Cuba's meager national budget and Cuba is examining what to do with them because of their lack of profitability. This situation represents a national security threat to the U.S. mainland, just by virtue of the existing capabilities of these laboratories and their stores of chemical and biological agents. What would it take to induce Castro to undertake covert biological and chemical attacks on Florida, New York City, or other U.S. locations? Or could Cuba currently be undertaking biowarfare experiments in the Caribbean Basin or in the U.S.? There are a lot of animal and plant diseases popping up in unexpected places around the world and in the U.S.

In view of the fact that there continues to be a global proliferation of biological weapons, such malicious experimentation cannot be ruled out. Cuba is of special interest because of its established capabilities, proximity to the U.S. mainland, and a history of supporting U.S. antagonists. It has been pointed out that Castro undertook a very effective past biological attack on the U.S. mainland just by emptying his hospitals and sending individuals with many kinds of communicable diseases to the U.S.

A PROPOSAL FOR ADDRESSING NATURAL AND TERRORIST THREATS TO FOOD AND AGRICULTURE

There should be a four-year-minimum appropriation of $350,000,000 for food and agriculture and critical infrastructure defenses systems development passed by the next Congress. This sum would be allocated not just to USDA and FDA, but to the development and training activities of state agriculture, state and local law enforcement, and schools of veterinary medicine. These funds could supplement resources already available for cross-training programs and combined surveillance, inspections, and intelligence programs in cooperation with the FBI, DOD, and the U.S. intelligence community.

The funded programs would address problem areas requiring special emphasis. The following suggestions are offered as potential major line items in a program budget:

- There should be a major emphasis on an integrated global human and agricultural health surveillance network, including information services for the public and for professionals involved in crisis and consequences management during food and agricultural emergencies.
- The surveillance network should be supplemented with an active research program on development of diagnostics methods and instrumentation and with development of some additional state diagnostic reference laboratories in collaboration with state universities. More reference laboratories for both human and agricultural diseases are known to be needed to ensure full coverage of epidemics on a nationwide basis. Diagnostic instrumentation is needed for use at medical and veterinary medical clinics, as well as for use in reference laboratories.
- There should be a specific emphasis on research on vaccines and pesticide research and development.
- There should be an emergency training program for first responders in food and agricultural emergencies, parallel to and associated with the DOD and FBI first-responder training programs.
- There should be a study component relating to review and improvement of incentives and enforcement tools available to the Secretary of Agriculture, FDA, and Department of Justice. This program would study creating positive incentives for safe food processing and developing more appropriate and varied enforcement tools for prosecuting biocrimes, product tampering, and various unsafe and questionable practices in agriculture. Training highly capable prosecutors for regulatory enforcement and prosecutions should be included in this activity.
- There should be a specific emphasis on expanding the open literature on biocrimes and bioterrorism and for funding special case studies of foreign-based human or agricultural epidemics for model development and assessment purposes. This task should include study of biowarfare research and development programs in effect today by world powers as they specifically build capabilities for attacking food and agricultural targets and associated infrastructures.
- A national panel of expert food and agricultural researchers and economic advisors should be appointed to advise the administration on how to prevent and minimize economic losses associated with natural crop and livestock diseases. This panel should include academic experts, state officials, and members of relevant associations.

- A Group-of-Eight conference should be organized to study methods for preventing or minimizing international transmission of infectious diseases and international circulation of dangerous pathogens. A global food and agricultural exports/imports safety convention should then be held. The conference agenda should include treatment of biocrimes and international agro-terrorism issues.

- At the same time, associations, agribusinesses, and pharmaceutical companies should be provided with cooperative private courses and workshops by law enforcement, defense, and intelligence agencies to educate and inform them about current and prospective threats and defense measures they can take. These courses and workshops should include protecting company facilities and critical infrastructures from chemical and biological terrorism and biocrimes. There is a lot of concern in corporate America about these threats, including threats to foreign-based assets and operations of American companies. However, most of these companies are taking no serious initiatives on their own. Federal and state leadership is needed here to show the way.

The estimated economic costs involved in natural epidemics of today and those that could result from large-scale attacks by sophisticated campaigns of irregular warfare are massive according to modeling studies. For that reason, cooperative action among the Office for Management and Budget, the National Security Council, and the Office of Science and Technology Policy is implied. We should launch a cooperative awareness promotion effort for inducing the Administration and Congress to seek the necessary money for the substantive four-year development program.

As for budget offsets for covering program costs, one possibility would be generating funds through reduced trade sanctions that are no longer justified. Demand for American farm exports and processed foods could then be increased to bring more income into farm households. A small portion of this increased exports income could go into building a fund for covering the proposed program expenses. This would be new money that could be earmarked for the proposed purposes, not an increased tax on existing income.

As for Administration support of this proposal, I would refer to the President's address to the United Nations in September 1998, which was dedicated to the issue of terrorism threats and meeting these challenges through international cooperation.[16] This speech presents a sufficient policy foundation to pursue the program outlined above. Few would argue that threats to food, agriculture, and associated infrastructures are threats of lesser significance than those presented to our embassies abroad in the long-term. In fact, hardening targets for conventional explosive attacks on buildings and airports may only increase the likelihood that the softer targets, such as food and agriculture targets, will be chosen by terrorists in the future. Remember that the most common target for biocrimes and terrorist attacks thus far has been food, not public facilities.

CONCLUSION

A comprehensive national defense system should provide for defenses from natural and terrorist threats to food, agriculture, and associated infrastructures. Improved linkages among agencies concerned with food and agriculture and those concerned with national defense and law enforcement are needed. These improvements require overcoming long-standing preferences for classifying information about terrorism-related events as secret and withholding it from the public. The basic thrusts of a food and agriculture defense program are outlined and are within the framework of the proposed $350 million four-year development program. This program can be operated in parallel with other current and proposed initiatives concerned with emerging and re-emerging infectious diseases.

REFERENCES

1. FREEH, L.J. May 1–3. Statement of Louis J. Freeh, Director, Federal Bureau of Investigation before the Senate Appropriation Committee Hearing on Counterterrorism. Available at: http://www.fbi.gov.
2. DEFENSE SCIENCE BOARD. 1997. 1997 Summer Study Task Force. DOD Responses to Transnational Threats. Volume 1. Office of the Under Secretary of Defense for Acquisition and Technology, Washington, D.C.
3. CARUS, W.S. 1998. Bioterrorism and Biocrimes: The Illicit Use of Biological Agents in the 20th Century: 5–9. Working Paper. Center for Counterproliferation Research. National Defense University. Washington D.C.
4. HICKSON, R.D. 1998. Logos subverted and the corrosion of illusionary liberations: A camouflaged strategic form of dialectical warfare and its infliction of mental sloth and hopelessness. Unpublished manuscript.
5. SOPKO, J.F. Winter 1996–1997. The changing proliferation threat. Foreign Policy: 105.
6. RICHMOND TIMES-DISPATCH. September 5. Japan edgy after poisonings: A4.
7. PURVER, R. 1995. Chemical and Biological Terrorism: The threat according to the open literature. Canadian Security Intelligence Service. Ottawa.
8. TAUXRE, R.V. 1997. Emerging foodborne diseases: An evolving public health challenge. Emerg. Infect. Dis. **3**(4): 425–434.
9. BEUCHAT, L.K. & J. RYU. 1997. Produce handling and processing practices. Emerg. Infect. Dis. **3**(4): 459–465.
10. HUXSOLL, D. 1993. The U.S. biological defense research program. *In* Biological Weapons: Weapons of the future. B. Roberts & G. Browder, Ed.: 58–67.
11. U.S. GOVERNMENT ACCOUNTING OFFICE. 1997. Combating Terrorism: Federal agencies' efforts to implement national policy and strategy. (GAO/NSIAD-97-254). Government Printing Office. Washington, D.C.
12. U.S. SENATE. Congr. Rec. July 15, 1998.
13. CARUS, W.S., op.cit.
14. CARUS, W.S., op.cit. pp. 131, 184–185.
15. CLINTON, W. 1998. Remarks by the President to the opening session of the 53rd United Nations General Assembly. United Nations. New York. (Available from: www.whitehouse.gov/WH/New/).

Agriculture and Food Security

FLOYD P. HORN[a] AND ROGER G. BREEZE

Agricultural Research Service, United States Department of Agriculture, 1400 Independence Avenue, SW, Jamie Whitten Federal Building, Washington, D.C., USA

There have been many other discussions about food and agricultural security, but all of these have been from the standpoint of abundance, availability, and quality. This volume is unique because it refers to three familiar terms frequently and openly: (1) agriculture, the science and art of farming; (2) biological warfare, the use of disease-spreading microorganisms, toxins, and pests against enemy armed forces or civilians; and (3) terrorism, the use of terror or violence to intimidate, subjugate, and demoralize, especially as a political weapon or policy.

Over the past ten years, there has been increasing awareness and public discussion of the increased risks to human health posed by highly infectious and dangerous viruses and bacteria, as represented by the lengthening list of new and emerging infections, from the Ebola virus to Lyme disease. But there has also been another less appreciated, but no less serious and growing threat—that of biological warfare and terrorism directed against armed forces and civilian populations. This has not been openly discussed to the extent it needs to be, perhaps from the fear of putting dangerous ideas into the heads of "bad people." This volume focuses on one aspect of the nation's vulnerability to bioterrorism and biological warfare that has been almost totally overlooked—American agriculture and our food supply system.

Threats to agriculture and our food supply system come from four sources: (1) individuals or groups with no specific goal, (2) individuals or groups with a specific goal, (3) sovereign states with established offensive biological weapons capabilities, and (4) sovereign states or groups to which these technologies are proliferating.

For example, individuals or small groups include U.S. citizens apprehended with large quantities of anthrax organisms in New Mexico (their motive was unclear) and ricin in Minnesota. Those possessing ricin were right-wing extremists targeting a deputy U.S. marshal and a sheriff. In Japan, there was a rash of "tampering" attacks on public food venues. Five people died and scores more became ill from cyanide or arsenic poisoning. The motives remain obscure but may be similar to past Tylenol tampering incidents in the United States. Oddly enough, in all aspects of terrorism, those with no defined agenda are becoming more common.

In contrast, the Aum Shinrikyo cult in Japan did have a goal: to precipitate an Armageddon from which its members would emerge as survivors. This cult made multiple biological attacks in Japan, one target was a U.S. military base, using *Coxiella burnetii* (the causative organism of Q fever), anthrax, and botulinum toxin, before releasing sarin and VX gases. Aum Shinrikyo continues to thrive and is popular with young scientists. Closer to home, more than ten years ago, a cult in Oregon caused

---

[a]Address correspondence to: Floyd P. Horn, Ph.D., Agricultural Research Service, U.S. Department of Agriculture, 1400 Independence Avenue, SW, Room 302A, Jamie Whitten Federal Building, Washington, D.C.; Telephone: 202-720-3656; Fax: 202-720-5427.

hundreds of *Salmonella* infections by contaminating a salad bar in a dispute with local authorities over zoning laws.

The Gulf War, the collapse of the former Soviet Union, and the end of apartheid revealed some surprising, well-established bioweapons capabilities in Iraq, Russia, and South Africa. Dispersal of those formerly employed in these programs, along with their knowledge and materials, have lead to proliferation of dangerous bioweapons opportunities for states and terrorist groups, particularly with bioweapons so relatively cheap and easy to create. There is growing awareness that given the overwhelming technological superiority of U.S. military forces, few nations would be willing to make a challenge on the battlefield, with or without biological weapons. But biological weapons could be employed for covert attacks on soft targets at home or U.S. interests overseas.

We have known for a long time that many naturally occurring microorganisms and toxins are potential biological weapons against people, animals, and crops. The idea is hardly new: the ten plagues of Egypt have been explained on the basis of naturally occurring pathogens and toxins. Smallpox- and plague-infected corpses were catapulted into besieged medieval cities by the Mongols and others. Smallpox-contaminated blankets were deliberately distributed to native populations in the Americas and elsewhere. The destruction of Ethiopian cattle herds by rinderpest virus in the late nineteenth century, outbreaks of glanders and anthrax in U.S. horses sent to Europe in World War I, and the various attempts to make weapons out of such agents as anthrax, foot-and-mouth disease, African swine fever viruses, and even the Colorado potato beetle, before and after World War II are other examples with an agricultural relevance.

So why are we apparently so suddenly concerned about the threat of biological weapons?

- Our knowledge of what has taken place in other countries since President Nixon halted U.S. offensive bioweapons research in 1969 is very limited.
- Investment in offensive bioweapons research has been substantial and sustained in certain countries.
- This knowledge is now proliferating to other countries and groups.
- Genetic engineering technologies introduced in the past 20 years give the capacity to enhance virulence of microorganisms and their toxins.
- The equipment and technology to replicate and deliver biological weapons are relatively simple and readily available.
- There are many naturally occurring microorganisms and toxins that are threats to agriculture and the food supply.

In this volume, Ken Alibek describes some of these issues. But one simple statistic of his former employer, the Soviet bioweapons research agency Biopreparat, gives us a useful perspective: Biopreparat had some 32,000 scientists and staff. The U.S. Department of Agriculture's Agricultural Research Service (ARS) has about 7,000 scientists and technical staff to address all agricultural research issues from air quality to catfish genetics. In his presentation, Alibek states that 10,000 of his employees were working solely on anti-agricultural bioweapons projects.

Why worry about bioweapons threats to agriculture and the food supply in the U.S.? A big part of the problem is that not enough people have been worrying. In fact, we should be very grateful to those few who have been. Colonel David Huxsoll, now at Louisiana State University's College of Veterinary Medicine and formerly the Commander of the U.S. Army Medical Research Institute of Infectious Diseases at Fort Detrick provided strong leadership in this area. More than ten years ago he began cooperative efforts with ARS at Plum Island on Rift Valley fever and Venezuelan equine encephalomyelitis vaccines.

Since the Gulf War, U.S. capacity to recognize and respond to biological weapons attacks has greatly improved. There have been many media reports of advances in battlefield detection and protection capabilities for our armed forces; increased emergency response preparedness for human disease outbreaks in major cities; and improved law enforcement, detection, and intelligence-gathering related to biological weapons.

These activities are some of the results of a series of Presidential Decision Directives relating to the threats of terrorism and biological weapons. Unfortunately, agriculture was not been included. Surprisingly, perhaps even incredibly, even the President's Commission on Critical Infrastructure Protection made no mention of agriculture in the discussion of domestic terrorism threats contained in its October 1997 report. Leadership on par with Dr. Huxsoll's has been far too meager.

Agriculture simply has not been a focus of our national attention in biological weapons preparedness, even though it is the foundation of our national security, the repository of our national wealth, the basis for our pre-eminence in the global marketplace, and the sustenance of our rural economy and ideological psyche.

- U.S. food and fiber system accounts for 13.% of gross domestic product (GDP) and for 16.9% of total employment.
- Exports alone are $140 billion and 860,000 jobs.
- America has "The most reliable, secure and safe supply of food at a reasonable cost that the world has ever known."
- Only about 2% of the population is involved in agriculture, leaving the remainder to engage in business, commerce, and the creation of wealth in other ways.

Agriculture in America is a lot more than these statistics. It is cowboys and vintners, and the runners in the pits at the Chicago Board of Trade. It is biotechnologists, florists, forest rangers, tenant farmers, rural cooperatives and the county fair. It is FFA and 4-H. It is human nutritionists giving us sound advice that could save us billions in health care costs and prevent suffering on a massive scale. Agriculture is a huge part of the American investment portfolio. It is an unequaled "jewel in the crown" of this great nation. It is our great concern that U.S. agricultural production, processing, and marketing system are more vulnerable than ever to deliberate assault by a wide range of biological warfare agents.

The first thing that probably comes to mind when thinking about a potential bioweapons attack on agriculture is that it would be aimed at eliminating some piece of the food supply so that part of the population would go hungry. This is a real risk for many countries today, as it has always been in the past. This is not the case for the

U.S., we have enough to feed ourselves and much of the world. We need to recognize that for us this is not a food issue. Nor is it just about money either. It is about terror. Terror, intense fear, is derived from the Latin verb *terrere*, meaning to frighten. Terrorism, as defined earlier by *Webster's Dictionary*, is "the use of terror or violence to intimidate or subjugate, especially as a political weapon or policy." In this regard a biological weapons attack on agriculture would be an attack on all civil society, on our economy, our way of life, our values, and our course and choices as a nation. It would be an attempt to frighten and intimidate the American public into a course of action they would not otherwise take.

We in the USDA will lead the national efforts to prevent this. For more than 100 years, the USDA has protected the public and a critical part of the U.S. economy from catastrophic risks related to agriculture and the safety of the nation's food supply. The USDA has no greater responsibility than to maintain this mission successfully in today's changing society and in the shadow of biological terrorism and warfare.

The USDA brings a great deal of experience to this new challenge. For more than a century, we have been helping farmers and ranchers find solutions to their problems, working to control animal and plant pathogens and pests on a daily basis. The critical federal role is to protect the public and our industries against pathogens and pests that might be introduced into the U.S. and spread rapidly through our plant and animal populations, including wildlife, fish, and birds. There are a very large number of these pathogens and pests, and not all of them are potential terrorist tools or biological weapons. They can be categorized by their characteristics:

(1) Pathogens that affect animals only (e.g., rinderpest virus)

(2) Pathogens that affect plants only (e.g., Karnal bunt of wheat)

(3) Zoonotic pathogens that affect animals and man (e.g., anthrax, rabies, and Brucella)

(4) Pathogens spread by insect vectors to animals and man (e.g., Venezuelan equine encephalomyelitis virus)

(5) Animal- and plant-related toxins (e.g., botulinum, ricin, aflatoxin, fumonisins, and tricothecenes)

(6) Advanced biochemical agents such as genetically manipulated organisms with enhanced toxicity or pathogenicity (e.g., baculoviruses incorporating the scorpion venom toxin gene) (This class has growing potential weapons relevance)

The zoonotic pathogens, those that infect animals and humans, are a troublesome group. Close cooperation with public health agencies will be necessary to control any outbreaks. These pathogens range from *Cryptosporidia*, not much of a disease problem in animal agriculture but an enormous public health risk, to Rift Valley fever, a serious problem for humans and agriculture.

Another problem is how to control diseases that affect some combination of humans, animals, wildlife, birds, and insect vectors. Rift Valley fever and Venezuelan equine encephalomyelitis are two examples of vector-borne diseases and both are known biological weapons agents.

If there has been a surprise in our uncovering of offensive bioweapons programs around the world, it is in the threats to agricultural crops. This should not have been a surprise at all—after all, every nation that has had, or still has, an offensive biological weapons program that includes components against animals and plants. We now know that certain laboratories targeted the major crops of North America (wheat, corn, barley, and soybeans) with rusts, bunts, scab diseases and more in forms that could be introduced into the U.S. with relative ease and with devastating results. Less threatening than an attack on public health, the impetus for such an attack could be purely economic, a situation exploiting the vulnerability of our commodities and futures markets and multiple soft targets at home.

Some of you may be familiar with Karnal bunt, a smut disease of wheat discovered originally in the Karnal region of India. *Tellita indica* is the causative organism. Typically, disease losses are not significant when tolerant cultivars are planted. There is no threat to public health from bunted wheat kernels. However, more than 30 countries, including the U.S. and many of our major trading partners, list *T. indica* as a quarantine pathogen.

In March 1996, Animal and Plant Health Inspection Service (APHIS) and ARS scientists identified samples of Karnal bunt–infected wheat kernels from Arizona. Subsequently, Karnal bunt was found in California and Texas. Had our international trading partners imposed a total U.S. quarantine, as was their right, our $6 billion wheat exports would have come to a standstill, with maximum disruption throughout many rural communities and our financial markets. However, because of the highly credible, rapid, and effective control and clean-up of the states concerned, our trading partners accepted a quarantine embracing only the affected areas as adequate, allowing our international trade in wheat from other regions to proceed. While no detailed cost-benefit analysis has been done to evaluate the quarantine and rapid-response actions, it is estimated that control and clean up by APHIS cost about $45 million and the maximum impact on exports was reduced to about $250 million, rather than $6 billion.

The recent experience with Karnal bunt illustrates the singular and unique vulnerability of U.S. agriculture to natural and deliberate attack by highly dangerous pathogens and pests that do not occur here and which would lead to international embargoes on our exports. The U.S. is so vulnerable because (1) it is the largest market in the world; (2) our animals and plant crops have little or no innate resistance to pathogens and pests that do not occur here; (3) there are very large populations at risk in small areas (2% of feedlots produce about 75% of U.S. beef); (4) the structure, size, and integration of the market promote rapid spread of pathogens and pests; and (5) increased international travel, tourism, and trade are multiplying the risks of pathogen and pest introduction.

For example, the distance from New York to San Francisco is the same as that from London to Baghdad. Yet it is by no means unusual to have foodstuffs from California on dinner plates in New York the day after harvesting. Or for that matter, tourists can be in China one day and back home on the ranch in the Imperial Valley the next.

Foot-and-mouth disease virus is the world's most important animal pathogen. The most infectious virus known, it can spread over 170 miles as an aerosol on the wind from an infected farm. The virus infects cloven-hoofed animals. In the United States, 100 million cattle, 70 million swine, 10 million sheep, and many of our 40 million wild animals are fully susceptible to any of the more than 70 different strains. One infected pig releases enough virus each day to infect, theoretically, 100 million cattle.

It has been estimated that a limited foot-and-mouth outbreak, that is one that affects about ten or so farms and is quickly diagnosed and eliminated, will cost the U.S. at least $2 billion. Anything more complicated and the costs skyrocket from there. Most of the costs come from the impact of international trade embargoes on animals and animal products. The U.S. swine industry is most vulnerable. The dramatic impact of this disease was recently seen in Taiwan, where the swine industry was destroyed. It should be noted that Taiwan was considered to have the best veterinary services in Asia.

One might argue that we have known of these risks for many years and have dealt with them very effectively within existing resources and strategies. Perhaps too effectively—for the most part the public is totally unaware of the risks and consequences and the USDA's behind-the-scenes efforts year-round and 24 hours a day, with diminishing resources. The American public has not seen funeral pyres of burning cattle in a foot-and-mouth disease outbreak in 70 years, and especially not since the advent of television. Television screens of other countries have been full of such images: in Britain for over a decade with the bovine spongiform encephalopathy epidemic, in Taiwan with foot-and-mouth disease, in Hong Kong with avian influenza, and in Europe with hog cholera.

We in the USDA want to sustain our record of success, but the new and growing threat of biological weapons demands that we find new ways and new resources to strengthen our ability to prevent and control these highly dangerous pathogens and pests.

The APHIS and the FSIS are the two USDA action agencies that would lead the response to any incidents and assure the purity, safety, wholesomeness, and integrity of the U.S. agricultural and food system. ARS is charged with providing them with the science and technological tools to do just that. The rest of this presentation outlines the six key steps needed to strengthen our defenses.

1. Prevent and Deter Terrorism within the U.S. and against U.S. Interests Abroad

The USDA must play an active role in the nation's counter-terrorism emergency preparedness plans and enhance intelligence, analysis, and sharing of information with other federal agencies, the states, local government, and the private sector. The USDA must reduce terrorist capabilities, such as unauthorized possession or importation of pathogens and pests. The USDA must protect government facilities and employees from attack and intimidation. We need to remember that terrorist threats are not all foreign. In the U.S., animal-rights extremists have attacked, ransacked and bombed, or burned ARS and APHIS facilities.

2. Maximize International Cooperation To Combat Terrorism

USDA must reduce and remove terrorist access to pathogens and pests that might be used as biological weapons. This can be accomplished through expanded cooperative efforts with foreign governments to control and eradicate foreign pathogens and pests at their sources. It must also participate in counter-proliferation programs to retain former Soviet weapons scientists in Russia to work on civilian projects. And the USDA needs to formulate accurate risk assessments.

Under the Freedom Support Act, ARS is participating in the Special American Business Internship Training (SABIT) Program for executives and scientists of the former Soviet Union who were engaged in weapons-related research. The knowledge, skills, and abilities of these scientists are being applied in the civilian sector to food safety, animal and plant infectious diseases, and biocontrol to benefit both the Russian and U.S. economies. Preventing emigration of these scientists to unfriendly states is a very high priority.

Commercialization of research, never a serious priority for these Soviet scientists in the past, is an integral part of the SABIT program. ARS is in many ways the most effective user of CRADAs with private industry in the whole of U.S. government, to fulfill the mission of the agency and to assure efficient transfer and commercialization of research and technology in the public interest. Recently, ARS signed our 800th agreement, indicating a program of about 14 CRADAs per 100 SYs, about five times the level of involvement found in the Department of Defense, the Department of Energy, and other agencies.

Currently, ARS has formed a team of microbiologists to explore with Russian scientists the potential for a commercial veterinary pharmaceuticals industry. This would not only create wealth in Russia to sustain a productive and beneficial science infrastructure, but would also facilitate better access by the U.S. to vaccines, drugs, diagnostics and other technologies necessary to deal with some of the threats we face.

3. Improve Domestic Crisis and Consequence Planning and Management

The USDA must (1) inform the public about the risks of unconventional weapons attacks on agriculture and the food supply system and the nature of the control measures that will be necessary; (2) enhance integration and coordination of crisis and consequence management, planning, training, command, and transition among federal agencies; and (3) improve our ability to respond to a domestic attack on agriculture and the food supply.

To accomplish this, we must increase our pathogen and pest interception capacity at U.S. ports of entry and within our county. The USDA relies heavily on its ability to prevent the entry of foreign pathogens and pests at ports of entry. Much depends on voluntary compliance by an informed citizenry. Deliberate smuggling of pathogens and pests bypasses this critical control point.

The USDA has crafted rapid-response plans to be lead by the APHIS and the FSIS. As a whole, the USDA must strengthen its human and physical infrastructure to protect against naturally occurring or deliberately spread pathogens and pests of plants and animals, including zoonotic organisms. We must also establish surge capacity to respond to operational and research needs. Current policies, procedures, and practices for epidemic pest and pathogen control must be reviewed paying special attention to changed public and media attitudes toward animal welfare, environmental quality, federal and state regulation, and the scientific bases for public policy. How information about agricultural pathogens and pest threats is made available to farmers, non-traditional customers, and the public must also be reviewed.

To meet the demands of our rapid-response plans, new tools will be needed, especially on-the-spot diagnostics that are accurate, quick, easy to use, and do not require biocontainment (because they don't use living organisms).

This does not mean that we will not need a national network of Biosafety Level (BL) 3 and 4 laboratories and animal challenge facilities. The Department of Defense still needs its BL 3 and 4 facilities at Walter Reed, USAMRIID, and the Navy Lab, even though there are more and more on-the-spot test methodologies. In fact, the work required to create these twenty-first century agricultural diagnostics and to devise the necessary vaccines will involve more work with virulent live microorganisms, some of which also infect people. For example, we found ourselves at the edge of comfortable safety practices in protecting ARS staff working with the recent avian influenza strain from Hong Kong. Fortunately, there is a drug that people can take to prevent this infection: without it, ARS could not have made the new vaccine and diagnostics for this emerging threat. Viruses like influenza that spread by aerosol are a real problem—for animals, birds, and our front-line staff. Despite the cost, we must expand and enhance our Biosafety Level 3 and 4 capabilities at Plum Island, Athens, and Ames. The USDA does not have any BL 4 capacity at this time and existing U.S. BL 4 labs could not handle farm species if there were a problem.

With regard to surge capacity, it is clear that the USDA needs to develop surge capacity for research, for diagnostic support, and for rapid-response operations. We need to establish a network of experts and resources before the need arises. We must draw on our land grant and other U.S. academic partners, the national labs, private industry, and others with something to offer.

4. Safeguard Public Safety and Protect Agriculture and the Nation's Food Supply System by Improving State, Local, and Private Capabilities

USDA must increase state, local, and private industry awareness and information-gathering capabilities regarding terrorist activities.

5. Safeguard Our Critical Infrastructures in Agriculture and the National Food Supply System

The President's Commission on Critical Infrastructure Protection did not include agriculture and the food supply system in its report. We need to work with industry and the states to rectify this omission.

We must identify and protect the cyber systems, including the global positioning system, and other critical infrastructures.

6. Research To Enhance Counter-Terrorism Capabilities

(1) The USDA must initiate new research programs to provide the tools to deter or detect, trace, and respond to an agricultural bioweapons attack in the twenty-first century because such risks will grow, not diminish, with time and technology changes. (2) The USDA must avoid technological surprise in weapons threats of biological origin, advanced biochemicals, and related activities resulting from current and future developments in foreign offensive agricultural bioweapons programs. (3) The USDA must work closely with the intelligence community in tracking agricultural bioweapons developments. In fact, intelligence analysis relating to the foreign or domestic development of advanced biochemical agents and genetically engineered organisms will be key to preventing technological surprise in this area.

These are major tasks. A host of microorganisms and toxins need to be mastered, and perhaps even more that have been genetically manipulated to enhance virulence. This three items summarize the 26 national programs that ARS employs to serve agriculture. Let me just mention a few items concealed in this innocuous list:

- On-the-spot diagnostics for all plant and animal pathogens and pests, including weaponized forms
- Epidemiologic mapping of pathogens and pests worldwide
- Genetically engineered vaccines that can be made without BL 3 or 4 biocontainment
- Mass vaccination systems for animals, fish, and poultry
- Alternatives to chemicals for mosquito and other vector control
- Alternatives to malathion and chemicals for plant pests
- Control of plant pathogens
- Genetically resistant crops
- New methods of animal pathogen control
- Hazard Analysis Critical Control Points plans for all disease threats
- New methods of euthanasia and carcass disposal
- Active research with foreign countries to control pathogens and pests
- Research to avoid technological surprise

CONCLUSION

We have laid the groundwork for rapid expansion to meet future challenges. We in the ARS believe we foresaw these threats ten years ago when the potential impact of genetic engineering technologies was brought to bear on the study of highly pathogenic viruses at Plum Island. With USAMRIID and others, we have made progress in these last ten years. With cooperation from federal, state, and local agencies, and the private sector, we can meet the challenges outlined in this article.

The Soviet Union's Anti-Agricultural Biological Weapons

KENNETH ALIBEK[a]
Battelle Memorial Institute, Arlington, Virginia 22202, USA

From 1975 until 1991, I developed biological weapons for the Soviet Union. From 1988 to 1991, I was the First Deputy Chief of Biopreparat, the civilian branch of the Soviet Union's anti-personnel biological weapons program. Although all of my work was conducted in the area of anti-personnel biological weapons, my duties included sitting on high-level scientific and administrative councils that oversaw the work of all of the branches of the Soviet Union's biological weapons program, including that concerned with anti-agricultural weapons. In particular, I attended meetings of the Military-Industrial Commission, which was responsible for governmental and administrative oversight of these programs. I was also a member of the Interagency Scientific and Technical Council, which met bimonthly to discuss scientific achievements and approaches in all areas of biological weapons development. This council's membership included Major-General Khudyakov, the director of the Ministry of Agriculture's Main Directorate of Scientific Research and Production Facilities. This directorate was responsible for the research, development, and production of anti-agricultural biological weapons. Through these meetings, I obtained a great deal of information on the Soviet Union's anti-agricultural biological weapons program.

Although the Soviet Union had been developing anti-personnel biological weapons since the late 1920s, it began to develop anti-agricultural biological weapons only in the late 1940s or early 1950s. By this time, the U.S. already had begun such work. My theory is that the Soviet Union developed an interest in anti-agricultural weapons specifically because the Americans were developing them.

The code name for this program was "Ecology," and as mentioned above, it was run by the Ministry of Agriculture's Main Directorate of Scientific Research and Production Facilities. The weapons developed by this agency can be divided into three categories: anti-crop, anti-livestock, and combined anti-personnel/anti-livestock. The anti-crop agents included wheat rust, rice blast, and rye blast; the anti-livestock weapons included African swine fever, rinderpest, and foot-and-mouth disease; and the combined anti-personnel/anti-livestock agents included anthrax and psittacosis.

Although some of the agents designated for weapons use in the anti-agricultural program coincide with those used in the anti-personnel program—anthrax being the most notable example—the Ministry of Agriculture independently developed its own production techniques. In contrast to the sophisticated reactor techniques devel-

---

[a]Address for correspondence: Dr. Kenneth Alibek, Battelle Memorial Institute, 1725 Jefferson Davis Highway, Suite 501, Arlington, Virginia 22202; Telephone: (703)413-7857; Fax: (703)413-3242.
e-mail: Alibekk@battelle-cc.org

oped to produce anti-personnel biological weapons, anti-agricultural weapons were generally produced by more primitive methods. For the anti-crop fungal diseases, this involved basic surface cultivation techniques. For anti-livestock weapons, cultivation generally involved live-animal techniques. It is worth noting that both of these basic techniques could be easily adapted by terrorists.

Also in contrast to anti-personnel biological weapons, anti-agricultural weapons were never produced on a regular basis or stockpiled. Instead, a number of facilities were equipped as mobilization capacities, to rapidly convert to weapons production should the need arise. The main facilities designated as mobilization capacities were located in Pokrov (a viral production facility) and in Almaty, Kazakhstan (a bacterial production facility). In addition, a facility at Vladimir and several labs in Galitsyno were involved in research, development, and pilot-scale production work. Finally, a facility at Otar, Kazakhstan (about 200 km from Almaty) served as a research, development, and testing facility.

By 1990, the Soviet Union had abandoned its anti-agricultural biological weapons program. The Ministry of Defense's philosophy of biological weapons use held that these weapons were for military use in case of total, global war. Anti-agricultural weapons, while suitable for terrorist use and particularly for disrupting the target country's economy, were not considered useful weapons in the global war scenario and therefore work on them was halted. It is interesting to note that the Soviet Union discontinued its anti-agricultural biological weapons program not as a result of any international pressure, but for purely strategic reasons.

However, it is important to bear in mind that many traditional anti-personnel biological agents, including some of those weaponized by the Soviets, would also have tremendous agricultural consequences. What might be considered the "classic" biological weapon, anthrax, is also fatal to livestock such as cattle, sheep, and goats. Brucellosis, which produces incapacitating illness in humans, can be lethal for livestock as well. Venezuelan equine encephalitis similarly would incapacitate humans but kill horses, as would glanders. In fact, I was informed by a colonel involved in biological weapons work for the Soviet Ministry of Defense that the Soviet Union had actually employed glanders biological weapons in Afghanistan. This would have had the dual effect of sickening the mujaheddin and killing their horses, which served as their main mode of transportation in the mountains.

In sum, we can learn a few lessons about anti-agricultural weapons from the Soviet Union's biological weapons program. First, these weapons were attractive enough to the Soviet Union that it spent considerable funds to develop them, and although shifts in Soviet strategy eventually resulted in the abandonment of anti-agricultural weapons, the potential economic and other consequences of these weapons may still attract other countries or groups. Second, simple techniques exist for the production of these weapons, which could be adapted for use by terrorist or other scientifically unsophisticated groups. And finally, in evaluating the potential consequences of an anti-personnel biological weapons attack, we must not overlook the concomitant effects of these weapons on livestock.

The Threat Posed by the Global Emergence of Livestock, Food-borne, and Zoonotic Pathogens

FREDERICK A. MURPHY

School of Veterinary Medicine, University of California, Davis, Davis, California 95616-8734, USA

INTRODUCTION

In the past few years, there has been an increase in emergent disease episodes, in the United States and globally. Given what we know about disease ecology, we can predict that such episodes will continue to increase in frequency and scale. Many of these episodes have involved livestock, food-borne and zoonotic pathogens. Many have presented a threat to human well-being and security. A few have involved what might be the ultimate threat of species jumping by the infectious agent, wherein transmission from an ongoing life cycle in an animal species leads to the establishment of a new life cycle in humans.

In general, there is no way to predict when or where the next important new pathogen will emerge; neither is there any way to reliably predict its ultimate importance. No one would have predicted the emergence of bovine spongiform encephalopathy (BSE) in cattle in the United Kingdom in 1986, and certainly no one would have predicted the species-jumping behavior of its etiologic agent, the BSE prion, causing new-variant Creutzfeldt-Jakob disease in humans. Given this reality, epidemiologic and laboratory investigations at the first sign of the emergence of a new disease are particularly important. The systems needed for such investigation are similar whether the threat is natural or malicious and whether the threat is to livestock, crops, or humans.

FACTORS CONTRIBUTING TO THE EMERGENCE OF NEW PATHOGENS

Many different factors can contribute to the emergence of new pathogens: microbial/virologic determinants (such as mutation, recombination, Darwinian selection), natural influences (such as ecological and environmental influences), and influences pertaining to human activity (such as societal, commercial, and malicious activities). Global human and livestock animal populations have continued to grow inexorably, bringing increasingly larger numbers of susceptibles into close contact and thereby favoring rapid transmission. A revolution in transportation has made it possible to circumnavigate the globe in less than the incubation period of most infectious diseases. Ecological changes brought about by human activity are occurring at a rapidly accelerating rate. Additionally, now, bioterroristic activities, supported by rogue governments as well as organized amateurs, are clearly increasing in scope and scale.

Ecological and Environmental Factors Favoring the Emergence of New Pathogens

One set of factors contributing to the emergence of new diseases relates to the capacity of microorganisms and viruses to adapt to extremely diverse and changing econiches. Arthropod-borne viruses are examples *par excellence* where emergence follows upon human actions:

- Population movements and the intrusion of humans and domestic animals into new arthropod habitats have resulted in many emergent disease episodes.
- Factors pertaining to unique environments and remote econiches (such as islands) have contributed to many emergent disease episodes.
- Deforestation has been the key to the exposure of farmers and domestic animals to new arthropods and the pathogens they carry.
- Increased long-distance travel of humans and transport of animals facilitates the movement around the world of exotic arthropod vectors and the pathogens they carry.
- New routings of long-distance bird migrations, brought about by new man-made water impoundments, represent an important yet untested new risk for introducing pathogens into new areas.
- Ecological factors pertaining to environmental pollution and uncontrolled urbanization are contributing to many emergent disease episodes.
- Ecological factors pertaining to water usage—that is, increasing irrigation and the expanding re-use of water—are becoming important factors in disease emergence.
- Global warming, affecting sea level, estuarine wetlands, fresh water swamps, and human habitation patterns, may be affecting vector-pathogen relationships throughout the tropics.

Behavioral, Societal, Commercial, and Iatrogenic Factors Favoring the Emergence of New Pathogens

There are many behavioral, societal, commercial, and iatrogenic factors that lead to the emergence of important new diseases or new disease patterns. We have overarching problems concerning (1) personal behavior and sexually transmitted diseases; (2) societal behavior and diseases associated with daycare; (3) community behavior and childhood diseases transmitted in the community; (4) medical care activities and hospital-acquired diseases, (5) medical care activities and diseases associated with immunosuppressive therapy and organ transplantation (and soon xenotransplantation); and (6) agricultural and commercial activities and diseases associated with food and water. This subject is very complex and diverse; food-borne diseases illustrate the sort of problems involved.

Food-borne Infectious Diseases

Recent public concern over food-borne disease is justified—the Centers for Disease Control and Prevention (CDC) estimates that there are approximately 40 million cases of food-borne disease and 9,000 deaths per year in the United States. This represents $6–9 billion dollars in economic loss per year. The great majority of food items that cause food-borne diseases are raw or undercooked foods of animal origin such as meat, milk, eggs, cheese, fish, or shellfish. The great majority of problems stem from microbial contamination. There are many different problems based on many different changes ongoing in the food industries:

- Changes in animal husbandry practices: the conditions under which food animals are raised have changed greatly and these often allow pathogens to enter the food chain at its source and to flourish, largely because of stress-related factors.
- Changes in sources of foods: we now import 30 billion tons of food a year, including meat, fruit, vegetables, and seafoods. Such foods represent an increasing proportion of our diet and often come from developing countries where hygiene and sanitation are less advanced.
- Changes in food processing and distribution systems: foods are produced by fewer but larger processors. This has resulted in increasing numbers of very large, interstate outbreaks of disease.
- Changes in food preferences and eating habits: there is a growing market for more ready-to-eat food products and novel, ethnic food products, which has created new situations where pathogens may be introduced into foods.
- Changes in the consumer population: there are increasing numbers of elderly and immunosuppressed persons who are at higher risk of severe food-borne disease. Also, consumers are now less educated in food handling compared to years ago.
- Changes in the food-borne pathogens themselves: new food-borne pathogens and more resistant pathogens have been identified as the causes of diseases that were unrecognized in earlier years.

Amazingly, specific foods causing large numbers of illnesses can escape recognition and critical control measures are often not identified. To deal with such problems, federal and state agencies and food industries are instituting programs that will (1) better identify important food-borne pathogens, (2) determine which foods are causing disease, (3) set in place better diagnostic tests, and (4) coordinate programs to deal with large interstate contamination episodes. This may seem simple enough, but in practice it is not—the epidemic of BSE in the United Kingdom is exemplary.

BSE was first diagnosed in the United Kingdom in 1986. As of August 1998 more than 174,000 cattle had been *reported* as infected. Modern statistical methods have indicated that the number of reported cases does not reflect the true scale of the epidemic: the true number of animals infected was about one million, roughly one half of which entered the human food chain in the United Kingdom. Epidemiologic evidence suggests that the epidemic was initiated by feeding cattle protein supplements derived from sheep meat and bone meal contaminated by the scrapie prion. Somehow, in a yet unknown way, the sheep prion seems to have become adapted to cattle, becoming the BSE prion. As the disease became established in cattle, it was amplified by inclusion in meat and bone meal of material derived from infected cattle. In

1988, feeding ruminant-derived protein supplements and bone meal was banned, but the ban was not well enforced. Even so, the epidemic had already been well established. But, the cattle disease epidemic is not the whole story.

In 1995, the BSE agent was reported to be the cause of a new human zoonotic disease, new-variant Creutzfeldt-Jakob disease (nv-CJD). By June of 1998, 27 cases had been reported in the U.K. and one in France. At first there was great skepticism that the cause of the human disease was BSE, but in the past year more and more evidence has been published that, taken as a whole, is overwhelming. In a recent report from The Royal Society, it is stated that "there is now a compelling case for regarding nv-CJD as the human manifestation of BSE."

At this point, but perhaps only with the wisdom of hindsight, it is being said that the ministry of agriculture and its veterinary regulatory agencies in the United Kingdom failed to react in timely fashion and with proper scope and scale of actions to deal with what was clearly a great risk to the livestock and related food industries of the country—every element of disease prevention/control responsibilities is being called into question. In 1998 the British government initiated yet another "BSE Inquiry," an amazingly large investigation into what went wrong. There are many lessons here, all pointing to the need for the very best epidemiologic and infectious disease expertise and resources, trained personnel ready to deal with the most novel of diseases as they emerge—whether this be an emergence in animals, crop plants, or humans. This must be the grounding for better public regulatory systems for intervening in the course of disease outbreaks. This, in turn, requires consideration for all steps along the "discovery-to-control continuum."

THE "DISCOVERY-TO-CONTROL CONTINUUM"

Given the reality of the threat, initial investigation at the first sign of the emergence of a new disease (whether it be in humans, animals, or crop plants) must focus on practical characteristics such as lethality, severity, transmissibility, and remote spread of the disease, all of which are important predictors of epidemic potential and overall risk. Clinical and pathologic observations and preliminary agent identification (the new realm of molecular biology) often provide early clues. The "discovery-to-control continuum" may be visualized as follows:

- Discovery, the recognition of a new disease in a new setting
- Epidemiologic field investigation
- Etiologic investigation
- Surveillance
- Diagnostics development
- Focused research
- Policy and marketplace matters
- Technology transfer

- Commercialization of some disease control elements
- Training and outreach
- Formalization of a lead disease control organization
- Control

The continuum starts with the perspicacious sphere of discovery, moves to the scientific area of risk assessment, and then on to the non-scientific, political area of risk management. Of course, not all of these elements are appropriate in every emerging disease episode—decisions must be made and priorities must be set: "We must do this, but we don't have to do that" and "What is the minimum that must be done to deal with this disease outbreak in this given circumstance?"

When one reflects on specific disease emergence episodes over the past few years, one is first struck by the importance of the scientific base upon which professional response depends. The base rests on laboratory and field investigation and the competence of those involved. The important early role in the continuum of local clinicians, pathologists (including medical examiners and forensic pathologists), and public health officials must be recognized. (Individuals are identified here in terms used in public health and human medicine—the parallels in veterinary medicine and crop agriculture are obvious.) The important early role of primary diagnostic laboratories and the reference laboratory networks that support them must also be recognized. Additionally, the importance of focused research must be recognized. In this era of the primacy of molecular biology, it bears reminding that many of the early investigative activities surrounding the identification of a possibly emergent disease must be carried out in the field, not in the laboratory. This is the world of "shoe-leather epidemiology."

What is involved in the intermediate phases in the "discovery-to-control continuum"? The continuum progresses to the general area of risk management, the area represented not by the question, "What's going on here?," but by the question, "What are we going to do about it?" This phase may include expansion of many elements: (1) policy and marketplace decisions and actions that involve the interface between government and the pharmaceutical industry; (2) technology transfer involving diagnostics development, vaccine and drug development, and medical/veterinary care activities and their adaptation to the circumstances of the locale where the disease is occurring; (3) commercialization, where appropriate, of diagnostics, vaccines, therapeutic agents or provision of alternate sources in quantities needed through non-governmental organizations or developed country government sources; (4) training, outreach, continuing education, and public education—each requiring professional expertise and each requiring adaptation to the special circumstances of the disease locale; (5) communications, at appropriate scope and scale, employing the technologies of the day, such as the Internet, and professional expertise (so often lacking).

More and more expensive, specialized expertise and resources come into play in the final phases of the continuum: (1) public health systems (including rapid case reporting systems, more sophisticated surveillance systems, vital records and disease registers, additional staffing and staff support, logistical support such as facilities, equipment, supplies and transport, legislation and regulation development, and senior management, administration and leadership); (2) special clinical systems (including isolation of cases by quarantine and/or strict barrier nursing, patient care, and in some cases an improvement in the general health of the population at risk); and (3) specialist and public infrastructure systems (including sanitation and sewerage, safe food and water supplies, environmental control, and reservoir host and vector control).

THE APPLICATION OF THE "DISCOVERY-TO-CONTROL CONTINUUM" TO BIOTERRORISM AND NATIONAL AND INTERNATIONAL SECURITY

Since a framework is needed to discuss "unnatural threats," and since it is a practical necessity to integrate all elements of our national capacity to identify, characterize, quantify, and respond to microbial and viral threats, it may be of interest to test the appropriateness of the "discovery-to-control continuum" with regard to the threat of bioterrorism.

First, discovery, or the recognition of a disease in a setting suspicious of bioterroristic activity: here, the elements of the continuum would likely be the same as in a natural threat episode. First responders and local clinicians, pathologists, and public health officials would most likely be the first to recognize an unusual disease episode. (Again, the public health and human medicine terms used here have obvious parallels in veterinary medicine and crop agriculture.) At first, it is unlikely that bioterrorism would be suspected—a terrorist-caused infectious disease episode would not present with the unique time/place characteristics as other kinds of terroristic actions. It seems likely that the usual civilian field-based epidemiologic investigation and laboratory-based etiologic investigation would follow, again because it is unlikely that bioterrorism would be suspected. Then, as the nature of the microorganism or virus was discovered, either because of unique clinical characteristics or more likely because of work in the primary laboratory or the state or national reference laboratory system, and as the magnitude of the episode became appreciated, the agencies concerned with national and international security would take over. At this point, quite a different continuum would be set into motion, but with the same purpose, i.e., disease control and prevention.

Given the character of infectious agents, including all those that have ever been considered as bioterrorism threats, a major focus of national security and law enforcement agencies would have to be microbial or viral source elimination, perimeter isolation, confinement and quarantine of exposed people and animals, area decontamination, and other area-based activities (although with human movement, with fomite transport, and with vector-borne agents, area confinement would not be simple). Seemingly, many of the professional activities called for are quite specialized and not readily available in any appropriate scale in the public health agencies. The question is whether they are available from the military or other national security agencies. This question must be answered in regard to supplies, equipment, on-site facilities, transport, and other logistics; staffing and staffing support; management and administration systems; trusted knowledgeable leadership; legislative and regulatory authority and law enforcement; special surveillance resources, case reporting, vital records disease register resources; and training and professional clinical outreach resources, public education resources, specialized patient care resources, technology transfer, etc.

The model for a national or even international bioterrorism response system is the national response system for dealing with a terrorist bomb or an earthquake or hurricane. Yet, responding to a bioterrorism episode is different enough, especially in regard to the need for specialized microbiological and virological expertise, equipment, supplies and logistics, that it would seem necessary to play out separate "war games" to answer questions about preparedness. Since the answer to the question of preparedness may in part be classified, it is not clear how much further this matter may be pursued.

NATIONAL RESEARCH FACILITIES AND PROGRAMS FOR DEALING WITH EMERGING PATHOGENS

From the public health standpoint, the question of facilities needs in this country must be dealt with—the same is certainly the case in regard to facilities for dealing with animal and crop plant pathogens.

What about high containment (BSL-3+ and BSL-4) laboratory facilities on the West Coast? Recent debate makes it clear that having two BSL-4 facilities in this country, at the Centers for Disease Control (CDC) in Atlanta and at USAMRIID in Frederick, Maryland (and one in Canada at the new center in Winnipeg) is not enough. There are plans for a few small BSL-4 labs in academic centers in this country and these may help in expanding basic research, but they will not support expanded surveillance, field-based research, and public health action research.

What about a new Plum Island Animal Disease Center (PIADC), with facilities equal to the best in the world? What about a new agricultural research/intervention program and facilities to deal with the threat of new, emerging crop plant pathogens?

CONCLUSIONS

So, who will be the world's public health doctor, the world's animal health doctor, the world's crop plant doctor, the world's environmental health doctor? And, where will these people work? [This usage taken from an Editorial in the *New York Times*, May 12, 1995.]

It seems that many institutions are saying that they have the answer to this question. In every case, the answer is in the form of proposals and massive funding requests to expand: (1) a global disease surveillance system, (2) a global diagnostics system, (3) a global integral research base, (4) a global communications system, (5) a global technology transfer system, (6) a global emergency response system, (7) a global training program, and (8) a global stable funding base.

New funding has been committed for some of these activities in the federal public health agencies (but not at the state level and not at the international level)—in my view, more is needed. If further funding is to be had, I believe that more, much more, public explanation of the importance of the issues must be forthcoming. In some cases, it seems that senior political staff officers are saying, "those fellows at CDC or at

PIADC just want more money to do their thing, to support their personal adventures," or "if things are so bad, then how did those fellows at CDC crack the hantavirus pulmonary syndrome episode in a week and how did they resolve the Ebola hemorrhagic fever episode in Kikwit in a month?" To deal with this kind of problem, on the most interdisciplinary level, here are two notions:

(1) Develop a greatly expanded communications system, providing comprehensive information to fuel the rising public expectations for more disease prevention and control action and the biomedical research that must underpin it.

(2) Integrate the public health sector's "national emerging infectious disease network" with networks focused on threats posed by livestock animal diseases, crop plant diseases, and bioterrorism and diseases pertaining to national and international security. The public should see such an overall "network of networks" as having a high benefit:cost ratio, a big bang-per-buck.

Contemporary Global Movement of Emerging Plant Diseases

RANAJIT BANDYOPADHYAY[a,b] AND RICHARD A. FREDERIKSEN[c,d]

[b]*Genetic Resources and Enhancement Program, International Crops Research Institute for the Semi-Arid Tropics (ICRISAT), Patancheru, A.P. 502 324, India*
[c]*Department of Plant Pathology and Microbiology, 120 L.F. Peterson Building, Texas A&M University, College Station, Texas 77843, USA*

ABSTRACT: Plant diseases are a significant constraint to agricultural productivity. Exotic plant diseases pose a continued threat to profitable agriculture in the United States. The extent of this threat has increased dramatically in the 1980s and 1990s due to the expansion of international trade in agricultural products and frequent movement of massive volume of people and goods across national boundaries. Introduction of new diseases has not only caused farm losses, but has also diminished export revenue since phytosanitary issues are linked to international commerce. Plant pathogens and their vectors have also moved across national boundaries, sometimes naturally and at other times influenced by the recent changes in trade practices. Sorghum ergot, Karnal bunt of wheat, potato late blight, and citrus tristeza are some of the most recent examples of enhanced importance of diseases due to the introduction of plant pathogens or vectors.

INTRODUCTION

The thread of agriculture passes through the fabric of all human endeavors. The United States Department of Agriculture (USDA) estimates that agriculture accounts for 13.5% of the national economy, and 17% of all jobs. Modern technology and innovative farmers have helped the United States to be a food-secure nation capable of offering food security in turn to several food-deficient nations around the world. A large proportion of Americans take agriculture and food for granted due to the abundant availability of quality food and a vibrant economy. Evolution of agriculture over the centuries has led to our dependence and survival on 30 principal food crops for nourishment.[1] Important among these are wheat, maize, rice, barley, soybeans, sugarcane, sorghum, potato, and oats, which contribute 85% of the estimated edible dry matter for humans. These crops also significantly contribute to export earnings. In the event of conspicuous losses in these crops, such as those experienced due to drought in Texas in 1998, disturbing consequences occur in several sectors of society and the economy, primarily in the rural sector. Modernization

---

[a]For correspondence: Dr. Bandyopadhyay: Phone: (650)833-6640 (California), 91(40)596161 (India); Fax: (650)833-6641 (California), 91(40)241239 (India).
  e-mail: r.bandyopadhyay@cgiar.org
[d]Phone: (409)845-1227; Fax: (409)845-6483.
  e-mail: r-frederiksen@tamu.edu

of agriculture, to increase efficiency and profitability and to meet the food needs of increasing population, has led to some profound changes in the way agriculture is practiced. Intensification of land use, modern cultivars, new land and crop management practices, changes in food preferences and associated food policy, dynamic trade policies, and frequent international movement of goods and people have all had effects on agriculture. These changes have varying effects on different components of the agricultural system.

The full potential of agriculture is rarely achieved due to the vagaries of nature, including losses caused by plant diseases. Plant diseases as a group causes substantial losses directly and indirectly by reducing the quantity and quality of food, feed, fiber, and industrial inputs. For example, losses due to soybean diseases alone were valued at $969 million in the United States in 1994.[2] Some plant diseases also affect the quality of the environment and aesthetics around us. Examples abound in world history of the catastrophic effects of crop losses due to plant diseases. The Irish famine in the last century and the Bengal famine in this century are just two examples of the devastation associated with plant diseases. Plant diseases have sometimes upset national economies, changed food habits, caused poisoning, transformed landscapes, and caused hardships in other ways.

Diseases are ubiquitous among plants, but some diseases are more damaging than others. Within a given agroecosystem, only a few diseases cause significant damage. The more damaging diseases are most often caused by infectious agents such as fungi, viruses, bacteria, and nematodes. Not all diseases of a plant occur everywhere the plant is grown. The causal agent of a disease must challenge a host cultivar at a stage when it is susceptible and at a time when environmental conditions are favorable to the pathogen. The relative importance of a particular disease on a crop is dynamic. In some cases, an important disease may become obscure when it is controlled with new disease management technologies. In other cases, an unimportant disease may reemerge as damaging with changes in agricultural practices, and still in other cases a new disease may emerge as important in a geographic area.

The spectrum of threatening diseases has changed dramatically in recent times. The concept of global movement of plant pathogens and the subsequent threat of exotic plant pathogens to agriculture is not new. Plant pathologists recognized the importance of this threat long before even Dutch elm disease was introduced to the United States around 1930. The American Phytopathological Society (APS) recognized the importance of emerging plant diseases in late 1920s and established the Committee on Investigations of Foreign Pests and Plant Diseases in the 1930s.[3] The committee recommended that federal agencies support studies on pathogens that are not found in the United States, but have devastating potential if introduced based on experiences of other countries. The Plant Disease Research Laboratory (renamed later as Foreign Diseases and Weeds Science Research Unit) of the USDA at Fort Detrick, Maryland was established to acquire knowledge on exotic plant diseases.[4]

Much later, the National Plant Pathology Board of the APS began a project on listing emerging and reemerging plant diseases throughout the U.S. The APS held a plenary session on these diseases in its 1996 annual meeting. Subsequently, the journal *Plant Disease* commissioned feature articles on some of these and other diseases. Readers are encouraged to read some of these excellent articles.[5–10]

Threats of new and reemerging diseases occur due to several factors. The major factors are: (1) movement of a new pathogen in a production system, (2) movement of one or more new virulent strains, or emergence of a new aggressive strain in an area where the pathogen existed, (3) introduction of new vectors that can transmit a pathogen efficiently, (4) changing agronomic practices that favor one or more components of epidemics of a specific disease, (5) increased pesticide use leading to development and proliferation of pesticide-resistant strains, (6) intensification of agriculture to maximize productivity and profit, (7) changes in cultivars, and (8) consistent change in climate in the short term. Some of these factors are intrinsic to crop husbandry, while others relate to extraneous, though important, forces such as trade, policy and international exchange. In this paper, we provide specific recent examples of enhanced importance of selected plant diseases that have linkages to introduction of new pathogens, movement of new strains of established pathogens, and vectors of plant virus.

INTRODUCTION OF A NEW PATHOGEN IN A PRODUCTION SYSTEM

Several exotic pathogens have gained entry in to the United States during the past decades. Some of the new entrants are economically insignificant while others have caused significant confusion and panic in the agricultural community. We provide examples of two recent entrants that have attracted considerable attention during the last three years.

Sorghum Ergot

Sorghum ergot is the most recent example of a disease that has spread rapidly in the Americas and Australia, taking the sorghum industry by surprise. Sorghum is the second most important feed grain in the United States, with grain valued at $2.2 billion. Sorghum is a staple food in several food-deficient countries in Africa and Asia. It is also one of the major export crops in the United States, earning almost $758 million in 1996.[11] In 1995, nearly 4.88 million metric tons of sorghum grain was exported to more than 22 countries.[11] It is estimated that the United States produces nearly 40% of world's hybrid seed valued at $435 million annually (A.B. Maunder, personal communication). Unfortunately, male-sterile lines used in F_1 hybrid seed production are highly vulnerable to ergot. Losses in seed production ranging from 10–100% have been reported from various parts of the world.[12] Commercial cultivation of hybrids is also vulnerable in cooler regions. Although sorghum ergot is an oldworld disease—it has been known to occur in Asia and Africa for more than 80 years—mention of its notoriety began with the introduction of hybrid seed technology in different countries. In spite of widespread use of hybrids in Australia, North America, and South America, ergot was of no consequence in these three continents due to the absence of the pathogen. That is no longer the case now.

The ergot pathogen attacks unfertilized ovaries to replace them with its own mass called sphacelia. Sphacelia exudes sweet, spore-laden, fluid, sticky honey dew. Later the fungal mass is converted into a sclerotium that contains potentially toxic alkaloids. Ergot causes crop loss by reducing the quantity and quality of seed, predisposing seeds to seedling diseases, and making harvesting and threshing difficult. Since the presence of ergot bodies increases the risk of disease transmission and toxicity, international trade of feed grain and seed are often jeopardized, as has been recently experienced with Mexico. An annual $5 million or larger increase in seed prices to producers have been projected due to new control practices to manage the disease. Additionally, potential trade implications show that every 5 cent/bushel decrease in sorghum prices costs the sorghum industry $31 million.

The global significance of the disease increased with the introduction of *Claviceps africana* in Brazil in 1995,[13] and in Australia in 1996.[14] Since then, the disease has been observed in rapid succession in other South American countries, Central America, the Caribbean, and North America. The disease spread was rapid within Brazil, Mexico,[15] the United States,[16] and Australia[12] resulting in its presence throughout the sorghum-growing areas in each country. In Brazil, ergot was found in several sorghum fields located within an 800,000 km^2 area a month after the disease was first officially recorded in the country. In Australia, the disease was confirmed in an area of 16,000 km^2 within a week of its first sighting, and after a month it had expanded to 70,000 km^2 around the locality of its first occurrence.

Ergot diffusion in South, Central, and North America and the Caribbean has been carefully monitored after the report of the disease in Brazil. By mid-1996, the disease had been recorded in Argentina, Bolivia, Paraguay, and Uruguay; by the end of 1996 in Colombia, Venezuela, and Honduras; and during the first quarter of 1997 in Puerto Rico, Haiti, the Dominican Republic, Jamaica, and Mexico. In late March 1997, ergot was observed in a sorghum field just north of the Rio Grande River near Progresso, Texas. By the end of 1997, the disease had spread throughout the sorghum-growing areas of the United States. Ergot's greatest threat to Mexico and the United States is in commercial hybrid seed production areas and in regions and situations where pre-flowering and flowering periods of sorghum extend into cool weather conditions.

The immediate impact of the ergot epidemic in Mexico and the United States was more important in the social and political aspects rather than in the agronomic. The farmers experienced the fear of the unknown. They did not know what ergot was or how it would affect their crops. The feed industry was concerned about the possibility of toxic alkaloids in the grain. Seed and feed grain exporters feared a possible shutdown of the export market in Mexico. In general, the incidence of ergot in the commercial grain sorghum fields was minimal in Mexico and the United States during the 1997 season. However, this incidence was higher than expected in some areas because the 1997 season was unique in terms of low temperatures and high relative humidity. Up to 80% incidence and 40% severity were recorded in some commercial hybrid fields in northern Tamaulipas, Mexico.[15] Growers, the seed and feed industry, researchers and extensionists have been trying to learn how to coexist with the disease and prevent it.

The ergot epidemic brought serious implications to the seed industry in Mexico and United States. Concerns about seed trade and regulation were first addressed. The seed production cost increased with the application of fungicides and additional sanitation procedures. Aerial applications of fungicides during the flowering stage are being used by the seed industry, but the efficient control of ergot has not been the best in some cases. Significant harvest problems were encountered in seed-production fields in southern Texas.

The sorghum commodity community in the U.S. met the ergot challenge remarkably well by sensitizing research administrators, policymakers, and the legislators to the importance of the disease on sorghum production and trade. Several universities, the National Sorghum Producers Association, the Agriculture Research Service, the Cooperative State Research, Education, and Extension Service (CSREES), the Foreign Agriculture Service of USDA, and state agriculture departments began funding ergot research in 1997. Research and extension activities began in several areas such as genetic resources, biology of the pathogen, epidemiology of the disease, toxicity studies, as well as control and management strategies involving phytosanitary issues, host plant resistance, chemical control, and pollen management. It is anticipated that considerable information will be available in the next few years to effectively manage and mitigate the threat of the disease in ergot endemic areas.

Karnal Bunt of Wheat

Karnal bunt of wheat is another disease that has received considerable international attention in the past few years due to issues related to quarantine and international trade. Initially restricted to India, Pakistan, Nepal, Afghanistan and Iraq, the disease was noticed in Mexico in 1972. The Animal and Plant Health Inspection Service (APHIS) began regulating wheat imports from countries with Karnal bunt in 1983 since presence of the disease in any country places it in considerable economic disadvantage from failure to export wheat. On March 8, 1996, the disease was officially recorded as present in the United States. On March 21, 1996, a "Declaration of Extraordinary Emergency" was announced by the Secretary of Agriculture to handle the disease and its repercussions. Intensive national surveys were carried out in 1996 and the disease was recorded in New Mexico, Texas, and California.[5] It is now recognized that Karnal bunt is neither a production nor quality constraint of wheat in the United States.[17] Therefore, surveys are planned again for 1998, but at 25% of the sampling rate of 1997.

Karnal bunt is a quarantine pathogen in most wheat importing countries due to limited global distribution of the disease. After the United States became a Karnal bunt–positive country, about 75 trade partner countries initially expressed concern in importing wheat from the United States. Several of these countries later allowed imports with additional phytosanitary declaration (that the grain came from an area free from Karnal bunt or where Karnal bunt is not known to occur) in good faith. As of February 1998, it was expected that almost $1 billion of the annual $6 billion wheat exports would be threatened due to curbs from importing nations because of the unresolved Karnal bunt crisis.[17] Domestically, Karnal bunt–affected areas were regulated at considerable economic cost. An economic analysis of regulation in the Imperial and Palo Verde Valleys of California showed that growers not only failed to realize nearly $77.6 million in profit from their inability to sell the grain, they also lost $57.75 million in production cost. The American Phytopathological Society, in a position statement on Karnal bunt, stated that the disease should not be regulated due to its limited agronomic consequence.[18] Nevertheless, surveillance and trade negotiations will continue to require considerable attention as long as importing nations have concerns about the disease.

MIGRATION OF NEW STRAINS

Late blight of potato is always mentioned when reviewing the impact of plant diseases on society and mankind. An epidemic of late blight in Ireland in 1845 led to famine, large-scale starvation, death, and population migration. This single event significantly changed the demography of Ireland and the United States. The late blight pathogen *Phytophthora infestans* also causes serious losses in tomato. Symptoms of the disease include blighting of the foliage resulting in defoliation and blighting of tubers (potato) and fruits (tomato). Under favorable conditions, the pathogen can spread rapidly causing severe defoliation and tuber spoilage. The pathogen survives in infected tubers and fruits and can sporulate readily on these substrates. The sexual survival structure oospore also helps overseasoning of the pathogen in parts of Europe and Mexico.

Late blight attracted the global attention again less than 150 years when it later reemerged as a devastating disease on potato and tomato worldwide. The primary reason for its reemergence in the United States is the migration of new strains of the pathogen. Extensive research[6] during the last 10 years has shown that before the 1980s a single clonal lineage of one mating type (A1) of the pathogen was dominant while the second mating type, A2, was rare or absent outside Mexico. Therefore, there was little diversity in the pathogen in the United States and Europe. Maximum diversity in the pathogen, including both mating types, occurred in the highlands of central Mexico where the pathogen and host co-evolved. Reports of the A2 mating type began to appear in the 1980s in Europe and in the 1990s in the United States and Canada. Subsequent analysis with markers such as mating type, allozyme genotype, and DNA fingerprints showed that the introduced isolates originated in Mexico. It is believed that export of potato tubers to Europe in the late 1970s and tomato fruit to the United States and Canada were responsible for long distance transport of the isolates.

Dramatic shifts in distribution of clonal lineages were noticed after the introduction of exotic strains. US-1 was the most dominant clonal lineage before the introduction of new strains. Since the early 1990s, three exotic lineages (US-6, US-7, and US-8) became increasingly important and replaced the resident strain US-1, which has not been found since 1993. US-6 and US-7 are pathogenic to both potato and tomato. US-6 became rare after 1993. The most infamous is the clonal lineage US-8. Its distribution expanded rapidly beginning with a single county in 1992 in northcentral New York, to most production regions in northeastern, southeastern, midwestern, western, and southern United States, including eastern Canada. Transcontinental shipment of potato seed tubers to distribute inocula far and wide, greater pathogenicity to foliage and tubers leading to faster establishment in a new field, and aerial spread of sporangia locally are some of the reasons for the widespread distribution of US-8. Introduction of US-8, a lineage with A2 mating type, has also created opportunities for sexual recombination with A1 mating type, thus increasing diversity in the pathogen as observed along the Pacific coast.[19] The US-8 lineage has other traits that make it dangerous and difficult to control. It has inherent resistance to metalaxyl, an effective curative fungicide, thereby making it difficult to control after the appearance of the disease. It is highly aggressive and requires more frequent sprays of protectant fungicides for disease suppression. US-8 also has unnecessary virulence genes, which suggest that it is likely to overcome newly deployed resistance genes for disease management.

The economic impact of potato late blight management has been quantified in the Columbia basin of Washington and Oregon.[20] Direct costs for late blight management, yield losses in the field, and losses in storage approached $30 million in the epidemic year 1995. Expenditure on increased frequency of fungicide use and their application alone was $25.3 million in 1995 compared to an estimated $6.6 million in 1994 when late blight was less severe. Indirect costs, such as processor costs, long-term revenue of storage facilities, etc., would further increase the losses from the disease. Psychological stress of growers and others experiencing losses is incalculable.

INTRODUCTION OF NEW VECTORS

Viruses are a major group of plant pathogens that cause serious economic losses. Vectors of different types transmit a majority of the plant viruses. Insects have been long recognized as a significant vector of plant viruses. Relationships between vector and virus vary. Some viruses are vectored by a single specific species with a high degree of specificity. However, a specific virus may also be vectored by several insect species, each varying in their efficiency to transmit the virus. Changes in distribution of specific vector species can have an impact on the economic significance of the disease caused by the virus.

Citrus Tristeza Virus

The example of Citrus brown aphid (BrCA) illustrates how introduction of an exotic vector can change the distribution of a disease from stabilized local endemics to extensive pandemics within a short period. Citrus is an important crop for the economies of several countries in South, Central, and North America and the Caribbean. Citrus tristeza virus (CTV) is one of the most economically important viral pathogens of citrus. The virus caused major declines in citrus production worldwide. Millions of citrus trees on sour orange rootstock have been killed by CTV in Argentina, Brazil, Spain, Venezuela, and the United States.[21] CTV proliferates mainly by graft transmission into new citrus growing areas. Several aphid species vector CTV in a semi-persistent manner and are responsible for secondary spread. Among these, BrCA (*Toxoptera citricida*) is the most efficient vector, followed by *Aphis gossypii* (cotton aphid). Other aphid vectors include *A. spiraecola*, *T. aurantii*, *A. craccivora*, and *Dactynotus jacae*. *Toxoptera citricida* has been shown to be at least 11 times more efficient in transmitting CTV compared to *A. gossypii*.[22] Destructiveness of CTV in South America during the 1930s and 1940s is linked to the combined presence of the virus with BrCA.[21] Similarly, devastation of the citrus industry in Venezuela during the 1980s can be linked to the introduction of BrCA. A severe decline-inducing (DI) strain of the virus was initially present in isolated areas of Venezuela. This strain of the virus was not a problem because the indigenous aphid vectors were not efficient vectors of the virus. The situation changed dramatically after the introduction of BrCA, which efficiently spread the severe strain extensively throughout the country. Due to the recent outbreaks of BrCA and the spread of severe strains of CTV, an estimated 185 million citrus trees are now highly vulnerable to CTV in the Caribbean Basin countries.[22]

CTV is not a new disease in the United States. It is present in Florida and California. Before 1996, *A. gossypii* was the major vector of CTV in both states. However, BrCA was detected in Florida in November 1995 in dooryards and small nurseries.[23] Since then, BrCA has spread rapidly in Florida despite an exhaustive eradication campaign. The U.S. citrus industry is now under a threat of CTV and the cause of the threat can be linked to the introduction of a new efficient vector.

CONCLUSIONS

Exotic plant diseases have caused appreciable losses in the agricultural sector during the 1990s. The threat from exotic plant diseases continues despite the recent spate of introduction of exotic plant diseases. The United States is still free from diseases such as tropical rust of corn and soybean rust, which are devastating diseases in countries where they occur now. Exotic pests remain exotic until they gain entry into a disease-free country. Not many diseases will remain exotic in the present climate of international trade and commerce. There is a need to stay prepared to combat the exotic diseases in case of their eventual entry into a new production system. Gaining knowledge to assess their risk and developing methods of control should receive priority at both the federal and state levels. Several universities and agencies of USDA have done a commendable job of responding to threats of emerging diseases after their entry into the United States. However, ability to conduct research on these diseases before entry of exotic diseases is seriously hampered by the lack of funding and interest from administrators and policy makers. This is because these diseases are not taken seriously until they enter the country and threaten agricultural productivity and trade. By then, much valuable time is lost. Guarding agriculture against anticipated threat of emerging plant diseases would require preparedness and an action plan based on sound research and policy issues.

Much similarity exists between factors that lead to emergence of diseases in plants and humans.[24] Evolving diseases in humans and the policies and practices to address them have justly received widespread attention.[25] With widespread consultations with several experts and associations, the Centers for Disease Control and Prevention (CDC) developed a prevention strategy for addressing emerging infectious disease threats in the United States. The strategy's four goals—surveillance and response, (applied) research, prevention and control, and strengthening infrastructure—apply to emerging diseases of plants too.

REFERENCES

1. HARLAN, J.R. 1995. The Living Fields. Cambridge University Press. New York.
2. WRATHER, J.A. *et al.* 1997. Soybean disease loss estimates for the top 10 soybean producing countries in 1994. Plant Dis. **81:** 107–110.
3. AMERICAN PHYTOPATHOLOGICAL SOCIETY. 1929–1936. Reports of the Committee on Investigations of Foreign Pests and Plant Diseases. Phytopathology **39:** 521–523, **40:** 459, **41:** 567, **42:** 481, **43:** 496, **44:** 570–571, **45:** 533–534, **46:** 497–498.
4. KINGSOLVER, C.H. *et al.* 1983. The threat of exotic plant pathogens to agriculture in the United States. Plant Dis. **67:** 595–600.
5. BONDE, M.R. *et al.* 1997. Karnal bunt of wheat. Plant Dis. **81:** 1370–1377.
6. FRY, W.E. & S.B. GOODWIN. 1997. Re-emergence of potato and tomato blight in the United States. Plant Dis. **81:** 1349–1357.
7. LATIN, R.X. & D.L. HOPKINS. 1995. Bacterial fruit blotch of watermelon: A hypothetical exam question becomes a reality. Plant Dis. **79:** 761–765.

8. LATTERELL, F.M. & A.E. ROSSI. 1983. Gray leaf spot of corn: a disease on the move. Plant Dis. **67:** 842–847.
9. MCMULLEN, M. *et al.* 1997. Scab of wheat and barley: a re-emerging disease of devastating impact. Plant Dis. **81:** 1340–1348.
10. POLSTON, J.E. & P.K. ANDERSON. 1997. The emergence of whitefly-transmitted geminiviruses in tomato in the Western Hemisphere. Plant Dis. **81:** 1358–1369.
11. FOOD AND AGRICULTURAL ORGANIZATION OF THE UNITED NATIONS (FAO). 1998. Production Statistics. FAO Web Page http://apps.fao.org
12. BANDYOPADHYAY, R. *et al.* 1998. Ergot: a new disease threat to sorghum in the Americas and Australia. Plant Dis. **82:** 356–367.
13. REIS, E.M. *et al.* 1996. First report in the Americas of sorghum ergot disease, caused by a pathogen diagnosed as *Claviceps africana*. Plant Dis. **80:** 463.
14. RYLEY, M.J. *et al.* 1996. Ergot on Sorghum spp. in Australia. Austral. Plant Pathol. **25:** 214.
15. TORRES, M.H. & G.N. MONTES. 1999. Sorghum ergot in Mexico. *In* Proceedings of the Global Conference on Ergot of Sorghum. C.R. Casela & J.A. Dahlberg, Eds.: 101–108. University of Nebraska. Lincoln, NE.
16. SAKEIT, T. *et al.* 1998. First report of sorghum ergot caused by *Claviceps africana* in the United States. Plant Dis. **82:** 592.
17. UNITED STATES DEPARTMENT OF AGRICULTURE. 1998. 1998 National Karnal bunt survey plan. On-line: http://www.aphis.usda.gov/oa/bunt/kbplan98.html
18. AMERICAN PHYTOPATHOLOGICAL SOCIETY. 1996. Position statement of the American Phytopathological Society. The use of quarantines for wheat Karnal bunt. APSnet. On-line: Karnal Bunt Symposium.
19. GOODWIN, S.B. *et al.* 1995. Direct detection of gene flow and probable sexual reproduction of *Phytophthora infestans* in northern North America. Phytopathology **85:** 473–479.
20. JOHNSON, D.A. *et al.* 1997. Potato late blight in the Columbia basin: an economic analysis. Plant Dis. **81:** 103–106.
21. ROCHA-PENA, M.A. *et al.* 1995. Citrus tristeza virus and its aphid vector *Toxoptera citricida*: threats to citrus production in the Caribbean and Central and North America. Plant Dis. **79:** 437–445.
22. YOKOMI, R.K. *et al.* 1995. Establishment of the brown citrus aphid *Toxoptera citricida* (Kirkaldy) (Homoptera:Aphididae) in Central America and the Caribbean basin and its transmission of citrus tristeza virus. J. Econ. Entomol. **87:** 1078–1085.
23. KNAPP, J.L. 1996. The brown citrus aphid: citrus tristeza virus relationship and management guidelines for Florida citrus. Citrus Industry **77:** 12–15.
24. ANDERSON, P.K. & F.J. MORALES. 1994. The emergence of new plant diseases: the case for insect-transmitted plant viruses. Ann. N.Y. Acad. Sci. **740:** 181–194
25. WILSON, M.E. 1994. Diseases in evolution. Ann. N.Y. Acad. Sci. **740:** 1–12.

Biological Warfare Training
Infectious Disease Outbreak Differentiation Criteria[a]

DONALD L. NOAH,[b,e] ANNETTE L. SOBEL,[c] STEPHEN M. OSTROFF,[d] AND JOHN A. KILDEW[b]

[b]*Air Force Medical Operations Agency/SGPA, 170 Luke Avenue, Suite 400, Bolling AFB, Washington, D.C. 20332-6188, USA*
[c]*Systems Research Center, Sandia National Laboratories, P.O. Box 5800, Albuquerque, New Mexico 87185, USA*
[d]*National Center for Infectious Diseases, Centers for Disease Control and Prevention, 1600 Clifton Road NE, Atlanta, Georgia 30333, USA*

ABSTRACT: The threat of biological terrorism and warfare may increase as the availability of weaponizable agents increase, the relative production costs of these agents decrease, and, most importantly, there exist terrorist groups willing to use them. Therefore, an important consideration during the current emphasis of heightened surveillance for emerging infectious diseases is the capability to differentiate between natural and intentional outbreaks. Certain attributes of a disease outbreak, while perhaps not pathognomic for a biological attack when considered singly, may in combination with other attributes provide convincing evidence for intentional causation. These potentially differentiating criteria include proportion of combatants at risk, temporal patterns of illness onset, number of cases, clinical presentation, strain/variant, economic impact, geographic location, morbidity/mortality, antimicrobial resistance patterns, seasonal distribution, zoonotic potential, residual infectivity/toxicity, prevention/therapeutic potential, route of exposure, weather/climate conditions, incubation period, and concurrence with belligerent activities of potential adversaries.

SCENARIO

You are the field medical commander in an air-transportable hospital deployed in support of an international peacekeeping mission. Responsible for the delivery of medical treatment and preventive medicine for more than 800 U.S. service personnel, your public health office is maintaining baseline incidence statistics for all categories of disease and nonbattle injuries. Over the past several days, your staff has witnessed an epidemic of acute respiratory illness among members of two flying squadrons. Although no deaths have yet occurred, the debilitating illness exhibits human-to-human transmissibility and significant resistance to first-line antimicrobials. In the face of this seriously degraded mission readiness, one of the belligerent nations steps up its military activity against the other.

[a]The views expressed in this article are those of the authors and should not be construed as an official Department of Defense or US Government position, policy, or decision.
[e]Address correspondence to: Major D. Noah, USAF, Armed Forces Medical Intelligence Center, Fort Detrick, Maryland 21702-5004.

At this point, you consider the potential for being on the receiving end of a biological warfare (BW) attack. You meet with your senior staff and ask the obvious questions: Is this a natural or intentional disease outbreak? How would we know if it were intentional? Are there features of intentional outbreaks that might allow us to correctly differentiate them from naturally occurring outbreaks?

The answers to these questions, unfortunately, depend upon many factors. Although obviously not all are readily identifiable, this paper will focus on biological warfare defense training as it pertains to the recognition of a biological attack.

As with the universal law of equal but opposing actions, the existence of BW agents in the hands of potential adversaries requires that various means of defense be employed. Designed to heighten the awareness of the populations at risk and hone their ability to respond while continuing their mission, these defensive means include health event surveillance and intelligence systems, vaccines, antitoxins, antimicrobial prophylaxis, agent detection and isolation devices, protective clothing, and training scenarios for decontamination teams, triage personnel, and field medical commanders.

The primary responsibility of the field medical commander is evaluation of alternative courses of action and selection of those with the highest probability of success (i.e., survivability). Within the chain of command, the field medical commander reports to the overall incident response commander (typically, non-medical). The primary functions of the field medical commander are: medical threat assessment; threat minimization through containment and decontamination; casualty triage; collection of relevant information; and analysis of projected courses of action. As the senior medical advisor, it might fall to the field medical commander to recognize an outbreak as either natural or intentional. This list of differentiating criteria may be useful to the field medical commander when faced with this dilemma. It is important to note, however, that each must be considered in context with the specific etiology and circumstances. In this presentation, we have also elected to highlight each criterion with individual examples rather than a single case scenario for the sake of flexibility and because of the difficulty of devising a hypothetical scenario that would appropriately utilize each criterion.

CRITERIA

1. Proportion of Combatants among Population at Risk

Continuation of the mission in the midst of a disease outbreak is dependent upon the health of the combatants and their immediate support personnel. An outbreak of chicken pox in the child care center, while demoralizing, likely will not degrade mission readiness. High morbidity due to colibacillosis or shigellosis among members of a deployed fighter squadron, on the other hand, may be a natural event that should justify higher levels of suspicion.

2. Temporal Patterns of Illness Onset (Epidemic Curve Characteristics)

A plot of the number of cases versus time defines the shape of the epidemic curve. A biological attack designed to degrade mission readiness will likely "explode" initially rather than display an insidious incidence. The possibilities of a combination of agents and temporal/spatial agent dispersion, however, underscore the need for fastidious sample collection and accurate, real-time intelligence reporting.

3. Number of Cases

Obviously, greater case loads would result in higher degrees of readiness degradation, demoralization, and support required for recovery. Although some natural infectious disease morbidity may be unavoidable, such as small numbers of cases of leishmaniasis among troops deployed to the Middle East, greater numbers among a discrete population might be suspicious.

4. Clinical Presentation

Many diseases have several possible manifestations and presentations. These depend upon factors such as the route of transmission and age and sex of susceptible persons. Anthrax, for example, most commonly manifests itself as a dermatopathy from direct inoculation from contaminated soil or animal products but can also present as a life-threatening respiratory disease after aerosol exposure.[1] Although cutaneous anthrax cases are routinely experienced in small numbers throughout many areas of the world, an outbreak of pulmonary anthrax would be cause for suspicion.

5. Strain/Variant

Viruses, bacteria, and rickettsiae are extremely dynamic microorganisms, capable of adapting to environmental changes. This adaptability means that even within a given species there is usually high genetic diversity. Our ability to fingerprint members of the same pathogenic genus and species is a crucial tool in both research and outbreak investigation. An example of this forensic application occurred when U.S. Centers for Disease Control and Prevention researchers responded to an outbreak of Ebola hemorrhagic fever in Kikwit, Zaire in May, 1995. Results of sequence analysis of viral isolates, which was available within days of recognition of the outbreak, showed that the etiologic viral variant differed by only three base pairs from the viral variant identified from the original outbreak in 1976 [personal communication, Dr. Ali S. Khan]. Although subsequent investigation determined that this outbreak was most likely a natural event, early concerns were raised about the possibility of an intentional introduction because of the genetic similarity of the strain.

6. Economic Impact

A pathogen does not necessarily have to affect humans to be an effective BW agent. For example, although the reintroduction of hog cholera in the United States would likely not cause any direct human deaths, the resultant economic devastation to the agricultural community would be practically immeasurable.[2] The recent introduction of foot-and-mouth disease into the Taiwanese hog population is a recent example of this economic devastation.[3]

7. Geographic Location

As mentioned earlier, diseases such as anthrax have geographic distributions that may serve as sentinel criteria. Ebola is an example recently made famous in fictionalized books and films; an outbreak in central Africa is considered a tragic resurgence of an endemic disease while even a small outbreak in the United States would be most surprising and suspicious. Modern military forces, however, are becoming increasingly mobile and interactive with indigenous populations. As a result, they have a correspondingly greater potential for translocating diseases to previously non-endemic areas through the unknowing transportation of infected vectors or people in incubation or clinical stages. Therefore, knowledge of the at-risk population's travel and contact histories may be essential in determining the etiology and likely source.

8. Morbidity/Mortality Rates

For most diseases, enough pathogenesis knowledge is available to generally predict how many people might be expected to get sick (attack rate) and, of those who become ill, how many are likely to die in a given outbreak. Significant variations from the expected may signal natural changes in organism virulence, changing human susceptibility, or intentional human manipulation. For example, influenza causes substantial annual morbidity and mortality in most areas of the world, with the biggest impact occurring in the elderly. Increases in either the number of cases, case fatality rates, or age groups affected may signal a pandemic caused by antigenic shift.[4] In 1918–1919, a global influenza pandemic was recognized, based largely on its unusually high case fatality rate, especially in young adults, which approached 10 to 20 per 1,000 worldwide.[5]

9. Antimicrobial Resistance Patterns

Resistance to antimicrobials is becoming more common as another manifestation of microorganisms' remarkable adaptive abilities. The patterns of resistance can be useful epidemiological markers for the origin of a particular infection. For example, an outbreak of tularemia among military personnel caused by a strain that is resistant to the standard therapeutic agents, while endemic tularemia infections are susceptible, might logically lead one to suspect the importation of a new bacterial strain or its intentional manipulation.

10. Seasonal Distribution

Coxiella burnetii, the causative agent of Q fever, is most commonly transmitted via the placental secretions from parturient sheep.[1] Therefore, this deadly disease is most commonly seen in the spring, during lambing season. An outbreak of Q fever in late summer or winter, or among personnel without sheep exposure, might not be easily explained as a natural occurrence.

11. Zoonotic Potential

Many pathogens, perhaps the majority, infect animals as well as man. These diseases are termed zoonoses.[6] The sudden occurrence of a zoonotic disease, such as brucellosis, in the absence of the natural animal host or reservoir and other likely sources of transmission, may be suggestive of an unnatural cause.

12. Residual Infectivity/Toxicity

When used on troops occupying friendly territory, BW agents are most useful when they incapacitate the opposition and then disappear. Some infectious agents, however, like *Bacillus anthracis*, form extremely durable spores when faced with adverse environmental conditions and may last for years in infected soil. The effectiveness and persistence of anthrax, therefore, may make it particularly suitable against opposing population centers creating long-term economic, political, and societal upheaval.

13. Prevention/Therapeutic Potential

Potential adversaries might be more likely to use BW agents: if they can immunize their own military personnel against the organism/toxin, or if the residual infectivity of the agent is limited. An outbreak of a disease among friendly forces, against which the adversary is known to possess a vaccine or antidote, would be considered highly suspicious, especially in the absence of likely natural causes.

14. Route of Exposure

Many diseases exhibit vastly different clinical presentations, periods of incubation, and mortality rates depending upon the route of transmission. The route of transmission of a naturally occurring incidence or outbreak can generally be predicted based upon the endemic form of the disease. Therefore, outbreaks due to an atypical route of transmission, particularly aerosol transmission, such as in the anthrax example above, are more suggestive of an intentional cause.

15. Weather/Climate Contributions

Weather factors, especially wind direction, temperature, and humidity, are important determinants of pathogen dissemination and disease occurrence and must be frequently taken into account in an outbreak investigation. A disease outbreak occurring downwind (or downriver) of a suspected BW agent production facility, such as in the Sverdlovsk anthrax disaster,[7–9] provides compelling evidence for an accidental or intentional release.

16. Incubation Period

Just as the clinical presentation may be a function of transmission route, so might the incubation period (the interval between exposure and illness onset). For instance, inhalation of bacteria or viruses may cause illness earlier than when the same pathogens are ingested or skin contacted. Incubation period is also strongly influenced by the inoculum or size of the infecting dose. A large number of people with tight clustering of incubation periods is suggestive of a high-dose, common exposure rather than sporadic cases with more varied dates of onset and periods of incubation.

17. Concurrence with Belligerent Activities of Potential Adversaries

Perhaps the most obvious sign of an intentional outbreak is one that occurs concurrent with the threatening actions of a belligerent force.

COMMENT

Current defense research among the world's potential belligerents likely includes biological weapons for a variety of important reasons. Today's conventional weapon technology is becoming increasingly complicated and costly. Conversely, biological agents are relatively easy to manipulate and inexpensive to produce and store. As treaty writers and verifiers also are discovering, innocent and necessary medical research requires the same scientific equipment and chemicals as does nefarious biological warfare agent production. Additionally, biological agents do not lend themselves to standoff detection and are difficult to trace back to the using group.

These facts should direct our attention toward a heightened awareness for, and an effective training effort against, the successful application of BW agents. Routine training leaves the individual soldier and operational unit prepared for anticipated employment tactics based on historical intelligence. However, shortfalls may emerge in the responsive decision-making required for future contingency operations. In addition to global surveillance initiatives, greater emphasis must be placed on honing data collection techniques, differential diagnostic tools and skills, and situational awareness skills of military health care professionals. One strategy includes the use of "no-notice" BW simulant attacks during Operational Readiness Exercises. These exercises provide an opportunity to assess medical readiness and responsiveness to a crisis situation immersed within a fully realistic mission scenario.

Finally, to be maximally prepared for natural or intentional disease outbreaks, military medical planners must become familiar with and exercise with local, state, and federal organizations also responsible for responding to such contingencies. Examples would include local and state health departments, the Federal Emergency Management Agency, and the U.S. Public Health Service.

The existence of biological warfare, and the recognition for greater awareness of that existence, is not new. We would do well to remember the words of Dr. Hans Zinsser:

> In the course of many years of preoccupation with infectious diseases, which has taken us alternately into the seats of biological warfare and into the laboratory, we have become increasingly impressed with the importance—almost entirely neglected by historians and sociologists—of the influence of these calamities upon the fate of nations, indeed upon the rise and fall of civilizations.[10]

ACKNOWLEDGMENTS

The authors thank Colonel Leo Cropper, Colonel Michael Parkinson, Dr. James Kvach, and Dr. Deborah Keimig for their sage editorial assistance.

REFERENCES

1. EVANS, A.S. & P.S. BRACHMAN, EDS. 1991. Bacterial Infections of Humans. New York. Plenum Medical Book Company.
2. WISE, G.H. 1981. Hog Cholera and Its Eradication. Publication APHIS 91-55. U.S. Department of Agriculture, Animal and Plant Health Inspection Service. Washington D.C.
3. TAIWAN CENTRAL NEWS AGENCY. 1997. BOFT struggles to find export markets for Taiwan pork. 17 April 1997.
4. NOBLE, G. 1982. Epidemiological and clinical aspects of influenza. *In* Basic and Applied Influenza Research. A.S. Beare, Ed.: 11–50. CRC Press. Boca Raton, FL.
5. WILLIAMS, R.J., N.J. COX, H.L. REGNERY, D.L. NOAH, A.S. KHAN, J.M. MILLER, G.B. COPLEY, J.S. ICE & J.A. WRIGHT. 1997. Meeting the challenge of emerging pathogens: The role of the United States Air Force in global influenza surveillance. Milit. Med. **162:** 82–86.
6. LAST, J.M., ED. 1988. A Dictionary of Epidemiology. Oxford University Press. New York.
7. U.S. DEFENSE INTELLIGENCE AGENCY. 1986. Soviet Biological Warfare Threat. Publication DST-1610F-057-86. Washington, D.C.
8. MARSHALL, E. 1988. Sverdlovsk: anthrax capital? Science **240:** 383–385.
9. MESELSON, M.S. 1988. The Biological Weapons Convention and the Sverdlovsk anthrax outbreak of 1979. FAS Public Interest Rep. **41:** 1–6.
10. ZINSSER, H. 1934. Rats, Lice and History. Little, Brown and Company. Boston, MA.

The U.S. Department of Agriculture Food Safety and Inspection Service's Activities in Assuring Biosecurity and Public Health Protection

BONNIE BUNTAIN[a] AND GEORGE BICKERTON

Animal Production Food Safety Staff, Washington, D.C. 20250, USA

Food Safety and Inspection Service (FSIS) is a regulatory agency within U.S. Department of Agriculture responsible for ensuring that meat, poultry, and egg products are safe, wholesome, and accurately labeled. The agency enforces the Federal Meat Inspection Act, the Poultry Products Inspection Act, and the Egg Products Inspection Act, all of which require federal inspection and regulation of meat, poultry, and egg products prepared for distribution in commerce for use as human food. In addition to its regulatory roles, FSIS acts in partnership with other federal and state governmental agencies, industry, consumer groups, extension services, researchers, and others to enhance communication networks, information sharing, and voluntary adoption of food-safety risk-reduction practices from farm to table.

In the animal production area, FSIS encourages improved collaboration among health officials responsible for animals and humans at the state and federal levels. The Interagency Coordinating Committee on Animal Production and Food Safety includes 15 agencies from four departments: the USDA, Department of Health and Human Services, Department of Defense, and the Environmental Protection Agency. The purpose of the Memorandum of Understanding among these agencies is to promote the sharing of information and to improve the coordination and collaboration of the member agencies' activities to protect human and food-animal health and safety. One of the key areas under discussion is animal-production monitoring and surveillance programs.

A key strategy of FSIS has been to encourage development of partnerships at the state level among animal and human health experts. This year FSIS funded efforts in ten state departments of agriculture to enhance their involvement and activities in live animal production food safety and to improve coordination and communication among state leaders in food safety and public health activities. An outcome of these partnerships will be better working relationships and networking among all stakeholders to respond to emergencies as well as routine animal health and food safety issues. In the area of consumer education, FSIS and the Under Secretary for Food Safety have played key roles in developing the national educational partnership

---

[a]Address correspondence to: Bonnie Buntain, Animal Production Food Safety Staff, 1400 Independence Avenue, SW, Room 0002 – South Building, Washington, D.C. 20250; Telephone: (202)690-2683; Fax: (202)720-8213.
 e-mail: bonnie.buntain@usda.gov

among industry, professional and consumer protection groups; educational organizations; and state, local, and federal agencies. With consumers more aware of food safety issues and microbial hazards and with key stakeholders working together to protect the food supply, our nation has important resources to draw on immediately to respond effectively to potential threats.

In emergency situations, regulations, policies, and procedures are in place to quickly remove unsafe meat, poultry, and egg products from commerce. FSIS Directive 8080.1 Rev. 2 (11-3-92) outlines the procedures for product recalls. These procedures are appropriate for use in cases of food that is adulterated for any reason, including terrorist acts. However, FSIS does not conduct any regulatory programs designated solely to combat terrorism.

The objective of FSIS recall policies, procedures, and actions is to quickly identify and remove from the marketplace hazardous and potentially hazardous meat, poultry, and egg products. The recall process plays a major role in the agency's commitment to improve food safety and better protect the public health. Accordingly, FSIS works in cooperation with federal, state, and local public health agencies, food producers, and various scientific and medical experts.

Since 1990, FSIS has conducted about 200 recalls of meat and poultry products that were not in compliance with the federal meat and poultry regulations. The recalls have accounted for more than 30 million pounds of product that had the potential of being a public health risk if consumed. (More detailed information on recalls since 1990 may be located at http://www.fsis.usda.gov/fsis/ophs/recalls/recalls.htm. Current recall information may be found at http://www.fsis.usda.gov/oa/news/yrecalls.htm#RNR.)

Recalls are crisis situations in which time is critical. Accordingly, it is FSIS strategy to perform the series of recall steps in an expedient manner to ensure that hazardous or potentially hazardous meat, poultry, or egg products are quickly removed from the marketplace. These steps are:

(1) Investigate consumer illness or injury to identify the cause.
(2) Coordinate investigative actions with federal, state, and local public health agencies and, when appropriate, assume a support role rather than a leading role.
(3) Determine the depth of the recall, that is, the distribution level at which the recall will be aimed.
(4) Advise appropriate federal, state, and local public health officials of recall action. When necessary, provide public notification to consumers.
(5) Conduct effectiveness checks to ensure proper return/disposition of recall product.
(6) Provide oversight of recall activities.
(7) Terminate recall action upon completion of recall activities.

If a recall is recommended following an extensive review of an incident, the Emergency Response Division (ERD) of the FSIS will establish the type of recall—Class I, II, or III—based on whether the non-complying product presents a real or potential health risk and the extent of product distribution in the marketplace. A Class I recall involves a strong likelihood that a product will cause serious adverse health consequences or death. For example, product contaminated with *E. coli*

O157:H7 bacterial pathogen that has reached retail stores normally would be subject to a Class I recall. A Class II recall involves a potential health hazard situation where there is a remote probability of adverse health consequences from the use of the product. A Class III recall involves a situation where the use of the product is not likely to cause adverse health consequences.

Recalls are voluntary actions by the manufacturer or distributor to remove noncomplying product from the marketplace. Recalls may be undertaken at any time by manufacturers or distributors on their own initiative or at the request of FSIS.

The ERD leads and coordinates all recall activities for the agency. Recalls are given priority attention at all levels of FSIS. Accordingly, ERD will investigate the facts and data of each incident and convene a Recall Committee to assess the possible existence of a public health risk. The Recall Committee comprises the ERD staff and other agency personnel knowledgeable of public health risks associated with meat, poultry, and egg products. For those illnesses or injuries considered unique or sensitive, the committee reports its findings to FSIS' Public Health Hazard Analysis Board, which determines the public health risk and advises the committee on recommended actions. The board is chaired by the director of the FSIS Epidemiology and Risk Assessment Division. Board members include FSIS personnel; representatives of other federal agencies, such as Centers for Disease Control, Environmental Protection Agency, and Food and Drug Administration; and subject-matter experts from the fields of medicine and academia.

ERD provides oversight of a recall action and ensures that all appropriate efforts are made for the return or proper disposal of the recall product. A recall action is terminated when proper disposition is made of the product and no further action is pending.

With regard to intergovernmental activities, FSIS and the Under Secretary for Food Safety serve on many interagency working groups to combat terrorism. On April 22, 1998, Secretary Glickman asked the Under Secretary for Food Safety to organize and coordinate an intradepartmental Food Emergency Rapid Response Team (FERRET) within the USDA. Other members of this group are the Under Secretary for Food, Nutrition, and Consumer Services; the Under Secretary for Farm and Foreign Agricultural Services; the Under Secretary for Research, Education, and Economics; the Assistant Secretary for Marketing and Regulatory Programs; the General Council, and the Inspector General. As appropriate, the group also will involve other senior officials of the USDA for advice and council. The group is charged by the secretary with preparing a plan to enhance the USDA's ability to respond rapidly to food safety emergencies and to develop a long-term strategy for preventing food safety emergencies.

On May 22, 1998, the Vice President announced a memorandum of understanding among the USDA, the Department of Health and Human Services (HHS), and the Environmental Protection Agency forming the Food-borne Outbreak Response Coordinating Group (FORCG). The goal of FORCG is to improve the response time to interstate outbreaks of food-borne illness among the responsible federal, state, and local agencies. To accomplish this mission, FORCG includes federal, state, and local agencies with outbreak response duties in the development of a national comprehensive and coordinated food-borne illness outbreak system. FORCG is co-chaired by the Under Secretary for Food Safety and the Assistant Secretary for Health.

Members of FERRET and FORCG recognize their roles in identifying and coordinating with state authorities responses to food-borne illness outbreaks. Therefore, the two groups of officials may be the first teams to receive reports of disparate outbreaks of food-borne illness that could result from the actions of bioterrorists. Leaders of these teams take appropriate actions in this regard.

In summary, FSIS and the Under Secretary for Food Safety have extensive formal and informal partnerships to coordinate monitoring, surveillance and outbreak-response activities. These networks can effectively provide the communication and coordination needed for emergency response among animal and human health and safety experts.

Safeguarding Production Agriculture and Natural Ecosystems against Biological Terrorism

A U.S. Department of Agriculture Emergency Response Framework

RON SEQUEIRA[a]

United States Department of Agriculture, Animal and Plant Health Inspection Service – Plant Protection and Quarantine, Center for Plant Health Science and Technology, North Carolina State University, Raleigh, North Carolina 27606-5202, USA

> ABSTRACT: Foreign pest introductions and outbreaks represent threats to agricultural productivity and ecosystems, and, thus, to the health and national security of the United States. It is advisable to identify relevant techniques and bring all appropriate strategies to bear on the problem of controlling accidentally and intentionally introduced pest outbreaks. Recent political shifts indicate that the U.S. may be at increased risk for biological terrorism. The existing emergency-response strategies of the Animal and Plant Health Inspection Services (APHIS) will evolve to expand activities in coordination with other emergency management agencies. APHIS will evolve its information superstructure to include extensive application of simulation models for forecasting, meteorological databases and analysis, systems analysis, geographic information systems, satellite image analysis, remote sensing, and the training of specialized cadres within the emergency-response framework capable of managing the necessary information processing and analysis. Finally, the threat of key pests ranked according to perceived risk will be assessed with mathematical models and "what-if" scenarios analyzed to determine impact and mitigation practices. An infrastructure will be maintained that periodically surveys ports and inland regions for the presence of exotic pest threats and will identify trend abnormalities. This survey and monitoring effort will include cooperation from industry groups, federal and state organizations, and academic institutions.

INTRODUCTION

Exotic pest introduction and outbreaks represent threats to agriculture, to our environment, and to the health and national security of the United States. It is thus advisable to identify relevant techniques and bring all rational approaches to bear on controlling introduced pest outbreaks. Recent political shifts indicate that the United States may be at increased risk for biological terrorism. Biological terrorism directed

---

[a]Address correspondence to: Ron Sequeira, United States Department of Agriculture, Animal and Plant Health Inspection Service – Plant Protection and Quarantine, Center for Plant Health Science and Technology, at North Carolina State University, Centennial Campus, Partners I Bldg., 1017 Main Campus Drive, Suite 2500, Raleigh, NC 27606-5202.

at agricultural production or natural ecosystems will have insidious effects on the productivity and ecology of the United States. Insidious because although not causing immediate human casualties, such activities threaten the supply of food, fiber, and alternative energy—thus a very real threat to our national security. Given the potential for devastating exotic species invasions, it behooves federal agencies to prepare information superstructures and train rapid-response cadres to become the first line of defense in case of biological terrorism. Careful planning, training of personnel, ground monitoring, remote sensing, use of forecast models, and spatial analysis using geographic information system (GIS)/global positioning system (GPS) technologies represent important elements in an integrated approach to the problem of pest detection and monitoring. This document outlines elements of the resource infrastructure and information superstructure that need to be formalized to improve agency preparedness to face threats to our agriculture and environment.

American agricultural production is central to assuring a safe supply of food. Despite the fact that Americans enjoy the world's most plentiful and productive agricultural systems, these systems are not immune to damage and even complete destruction by exotic species. Beyond agricultural production, the natural environment is equally challenged by invasive pests. American history is filled with anecdotes of the disastrous effects of invading diseases and insects. In 1904, an epidemic known as the "chestnut blight" caused by an Asian fungal agent, *Endothia parasitica*, resulted in the near extinction of the American chestnut. The destruction of vast expanses of chestnut worsened the effects of the Great Depression in the southeastern United States. The American chestnut had provided durable and decay-resistant lumber for furniture production. Its bark fostered the tanning industry and farmers pastured their hogs in native stands for the nutrition-rich nuts. The blight was not eradicated and chestnut stands never recovered. Largely due to the economic, ecological, and social impact of the demise of the American chestnut and two other tree diseases, the Plant Quarantine Act of 1912 was passed by the U.S. Congress, resulting in the first national regulatory quarantines in the U.S. [The Plant Quarantine Act of 1912 as well as other regulations subsequently passed by Congress authorizes the USDA to take the lead in responding to threats to animal and plant health.] Another example deeply rooted in the folklore of the American South is the nineteenth century introduction of the boll weevil (*Anthonomus grandis*), which devastated cotton production and continues to be the subject of intense eradication efforts. Current examples (1997–1998) include the multi-billion dollar threats posed by the Mediterranean fruit fly (*Ceratitis capitata*) and citrus canker (*Xanthomonas campestris*) to the fruit and vegetable production in several southern states; and the nearly completed campaign against wheat karnal bunt (*Tilletia indica*).

These examples are most likely the result of accidental invasions. Given that unintentional introductions of exotic pests do occur, the U.S. Department of Agriculture–Animal Plant Health Inspection Service (USDA-APHIS) has developed plans to face these threats. However, biological terrorism directed at agricultural production poses different challenges.

Intentional introductions of exotic invasive species may differ from accidental introductions in the following ways: (1) use of non-traditional pathways, (2) increase of the probability of survival of the pest in-transit, (3) widespread dissemination of the disease from disparate foci, (4) use of highly virulent strains, (5) high rates of inoculum, (6) introduction into remote areas, (7) targeting of susceptible production areas, (8) targeting of susceptible natural environments, (9) release of multiple species simultaneously, and (10) precise timing of releases to coincide with maximal colonization potential. Current response structures were not developed to address such possibilities. Importantly, the expanded trade and movement of people as a result of the "globalization" of the economy have already taxed the existing USDA structures and tested our resources.

USDA-APHIS battles invasive species along multiple fronts at any given period. During the 1997–1998 period, no less than 20 introduced plant pests (usually near ports of entry) were the targets of eradication campaigns, strict quarantines, or regulatory actions by the USDA. These included citrus canker (*Xanthomonas campestris*), Mediterranean fruit fly (*Ceratitis capitata*), boll weevil (*Anthomus grandis*), silverleaf whitefly (*Bemisia argentifolia*), pink hibiscus mealybug (*Maconellicocus hirsutus*), Mexican fruit fly (*Anastrepha ludens*), Caribbean fruit fly (*Anastrepha suspensa*), Asian long-horn beetle (*Anoplophora glabripennis*), gypsy moth (*Lymantria dispar*), pink bollworm (*Pectinophora gossypiella*), imported red fire ant (*Solenopsis invicta*), japanese beetle (*Popillia japonica*), karnal bunt of wheat (*Tilletia indica*), bovine screwworm (*Cochliomya hominivorax*), and chrysanthemum white rust. Additionally, several animal diseases were targeted for management or eradication.

EXISTING USDA EMERGENCY RESPONSE STRUCTURES

The regulatory branch of the United States Department of Agriculture is APHIS. In this capacity APHIS' mission includes safeguarding American agriculture productivity and the natural environment. Emergency-activities responsibility resides within the Plant Protection and Quarantine (PPQ) and Veterinary Services (VS) sections of USDA-APHIS.

APHIS-PPQ and APHIS-VS have developed independent emergency-response procedures and structures. The National Plant Board, an organization of the state plant pest regulatory agencies in the U.S. and Puerto Rico, coordinates emergency activities at the state level.[1] Both the USDA and the National Plant Board (for plant-related emergencies) coordinate joint state and federal actions. APHIS-VS has developed a formal organization, the Regional Emergency Animal Disease Eradication Organization (READEO) and codified its operations in a manual.[2] APHIS-PPQ has an analogous emergency-response system codified in an operations manual.[3] FIGURES 1–3 show the functional and structural organization of APHIS-VS and APHIS-PPQ emergency-response operations. The discussions in this paper emphasize the "plant" side, reflecting the expertise of the authors, but similar issues are applicable to the "animal" side of APHIS emergency operations. The PPQ emergency procedures manual describes the sequence of events, the procedures, regulatory authority, structure, and roles and responsibilities during an emergency program.

Briefly, emergency procedures are triggered when the New Pest Advisory Group (NPAG, a formal coordinating team at the Center for Plant Health Science and Technology-USDA-APHIS-PPQ) receives notification that a plant pest has been discovered in the United States. The NPAG convenes a panel of experts on the specific pest

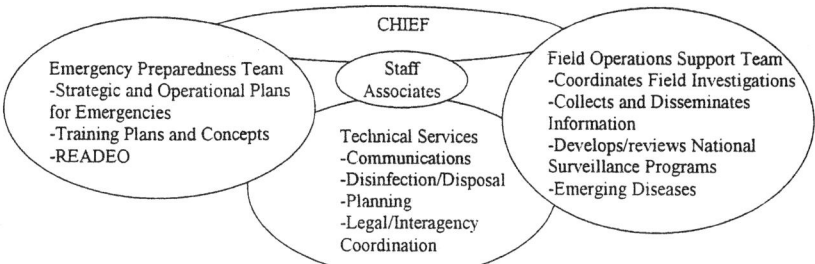

FIGURE 1. Emergency Programs Staff: Functional Areas. USDA-APHIS-Veterinary Services.

group and submits a recommendation to the PPQ administration. This recommendation may include actions such as: continue monitoring and survey, do nothing, commence eradication and declare quarantines and regulatory containment measures. When regulatory action is deemed necessary, the Secretary of Agriculture is authorized to do the following: establish or modify quarantines and regulations to carry out emergency eradication programs against new plant pests; restrict and prohibit the entry and interstate movement of plants and products to prevent the entry and interstate spread of pests; declare an extraordinary emergency when a new plant pest is present in the U.S. and threatens the agriculture of the United States and state measures are determined inadequate; and cooperate with states, farmers, industry groups, associations, and other countries of the Western Hemisphere to carry out operations to control or eradicate pests posing a significant hazard or threat to the United States.

Broad authorities to support emergency programs include those stated in the following Congressional Acts: Federal Plant Pest Act, approved May 23, as amended; the Plant Quarantine Act of 1912, as amended; The Organic Act of 1944, as amended; Federal Noxious Weed Act of 1974, approved January 1975; Cooperation with State agencies in the administration and enforcement of certain Federal Laws Act, approved Sept. 1963; Federal Insecticide, Fungicide, and Rodenticide Act, as amended; The Honeybee Act; Joint Resolution of April 1937, as amended, and the Golden Nematode Act, approved June 1948.

If an eradication program is deemed necessary, the eradication procedures begin by establishing a federal/state cooperative structure that includes the following functional areas: air operations, regulatory, survey, safety, and control, among others. Each of these areas is supervised by a section "officer." One of the first priorities during an emergency is to conduct surveys and delimit the extent of the infestation or epidemic. In conjunction with this activity, the regulatory section will implement quarantines, coordinate enforcement with local agencies, and monitor movement of susceptible materials. The control section determines which strategies are most appropriate for containment and eradication and coordinates the implementation of these strategies. If air operations (e.g., aerial bait sprays) are necessary then the control officer works in close coordination with the air operations officer and local airports and transportation authorities. Environmental safety is assured by a team of

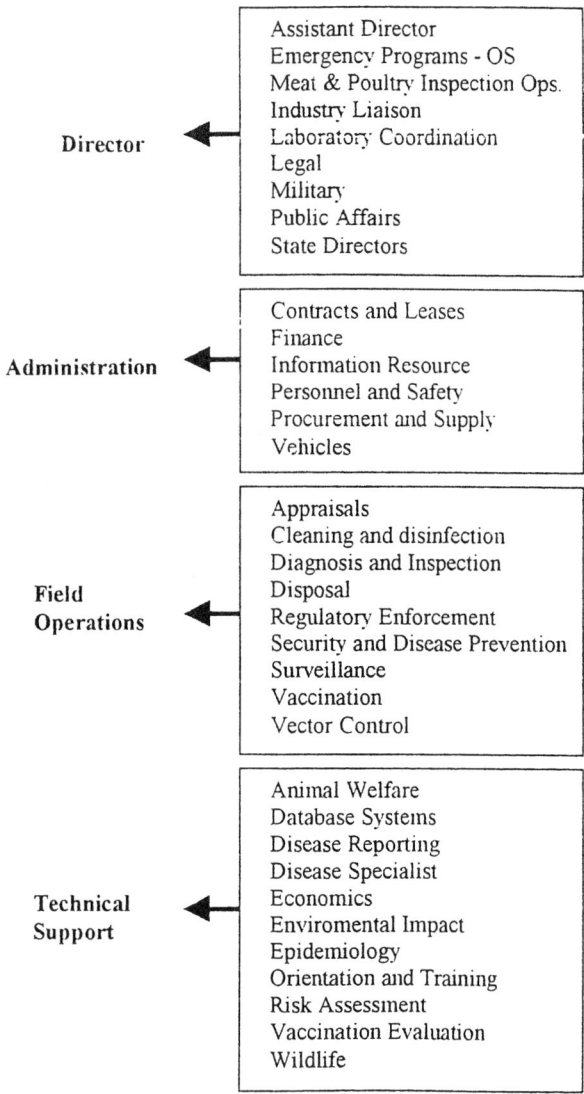

FIGURE 2. Emergency Programs: READEO Structure. USDA-APHIS-Veterinary Services.

environmental monitoring experts working under EPA and APHIS guidelines to ensure compliance with regulatory mandates, when applicable.

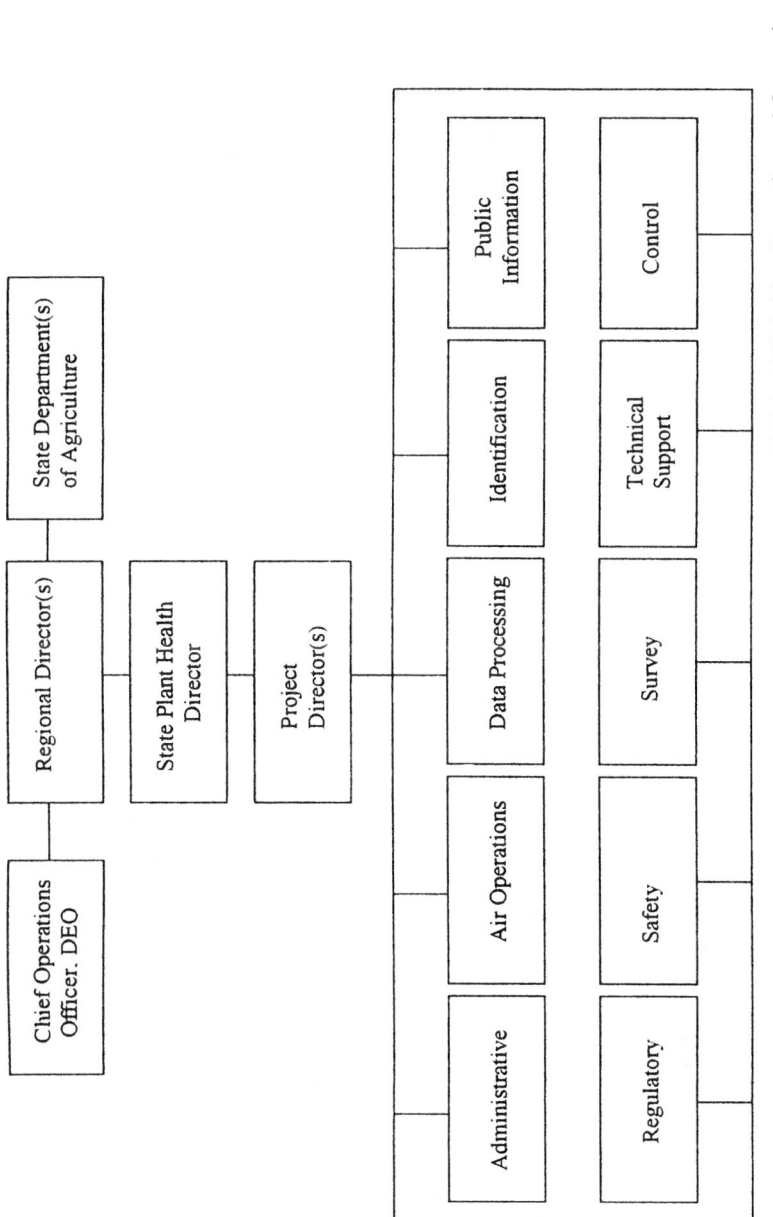

FIGURE 3. Emergency Programs: Federal/State Emergency Projects—USDA-APHIS. Plant Protection and Quarantine.

RISK ASSESSMENTS AND THEIR ROLE IN DEFENDING THE UNITED STATES AGAINST INVASIVE SPECIES: A CASE STUDY

The example developed here has a twofold purpose. First, it aims to illustrate the importance of risk assessments in safeguarding U.S. agriculture. Second, it illustrates how some recent information management, monitoring tools (remote sensing, earth-observing satellite imagery, meteorology), GIS, and simulation models can be brought to bear on this element of countering biological terrorism. The case study examines risk assessment for a disease of wheat, karnal bunt (*Tilletia indica*). It is proposed here that a key element in countering the threat of bioterrorism is knowledge of the risk posed to production areas by key exotic species. Risk assessments and simulation models are key elements in this process.

This case study is not an example of bioterrorism. Rather it is a case that illustrates methodological approaches relevant to agriculture security. The implementation of agricultural security measures will require structural changes within the USDA.

The Assessment of Risk

The assessment of risk as defined in regulatory agencies implies the identification of the risk factor, the estimation of the probability of occurrence, and the evaluation of the outcome.[4–11] One of the key tasks when dealing with a new pest is to determine the likelihood that it will establish within agricultural or natural environments in the United States. The probability of this establishment depends on several factors: that the organism in question is on a pathway leading to production regions; the probability that the organism will survive in transit; the probability that the organism establishes a reproductive population where introduced; and the probability that the organism will spread beyond the initial colonization area. The overall risk model utility is given by: P[establishment] × Outcome from Establishment. The probability of establishment for non-indigenous pests has been characterized by Orr and colleagues[12] as:

P[establishment] =
  $f(P$[pest in pathway to host], P[entry&survival], P[colonization], P[spread])

A risk assessment is often a probabilistic expression of the chance of an unwanted occurrence; in our case the negative event is the establishment/spread of a given pest in the United States. The results of a risk assessment are often expressed using probabilistic expressions or comparative distributions of families of stochastic variables. Given the intended use of the results of typical assessments (i.e., their use by panels of administrative/regulatory officials and the communication to lay public) the communication of the risk for a given pest is best expressed in straightforward categories (such as high, medium, and low) or other categories easy to relate to well-known experiences.

Case Background

USDA inspectors recorded infections of karnal bunt (*Tilletia indica*) in March 1996 from seed lots of durum wheat in Arizona. The quarantine response, nationwide surveys, trade impacts and historical information have been reviewed and are maintained on a public website maintained by USDA-APHIS at http://www.aphis.usda.gov/. A recent review of literature[13] lists key scientific references. Only details that have specific relevance to the current objectives are reviewed here. The objective of this section is to describe an approach to assess the risk to wheat posed by karnal bunt and to communicate this assessed risk by categorizing all United States production areas into zones of high, medium, and low risk.

Biology of Karnal Bunt

The life cycle for *T. indica* is complex and requires a soil-borne phase (teliospore) in which germination occurs at 20–25°C or 15–22°C (alternating light versus *in vitro* conditions). Teliospores germinate near the soil surface producing a promycelium, which in turn bears primary sporidia. These primary sporidia produce new spores (allantoid and filiform-like), which infect the inflorescence or developing grain. Teliospores have been reported to be resilient to temperature extremes and to germinate at a wide range of temperatures (5–25°C). Spores infecting wheat reproductive organs are similarly active over a wide range, but with different, and narrower, temperature and moisture regimes than those reported for teliospores. Under optimal humidity these spores may germinate at temperatures from 8–20°C. Temperatures greater than 20°C are cited as unfavorable.[13–16] Jhorar and colleagues conducted field experiments and concluded that optimal conditions in the field were somewhat different than those reported for controlled chamber studies. They summarized their results in a linear model: DI = −0.8 + 1.5 HTI, where HTI = ERH/TMX. (DI is disease index, HTI is a humid thermal index, ERH is evening relative humidity, and TMX is temperature maximum.)

Thus the model's HTI uses maximum observed temperature and relative humidity to predict establishment of *T. indica* assuming that a susceptible host is present. Jhorar and colleagues[17] found that an HTI of 2.2–3.3 favored the disease.

U.S. Production Characteristics Relevant to Establishment Likelihood

Wheat is the leading U.S. export crop. Wheat varieties grown in the United States include more than 200 commercial cultivars. However, only a small proportion is responsible for most of the wheat sown. Wheat is planted in both winter and spring. Most (>70%) of all wheat is sown as winter wheat. Several agronomic types or varieties of wheat are grown and are categorized according to factors such as milling characteristics, where they are grown, and when they are planted. Wheat types include: red, white, winter, durum, etc. (published on websites of the National Association of Wheat Growers and the USDA NAS).

A key production feature for this risk assessment is the fact that wheat is planted during given "sowing" windows, which vary by region. These sowing windows show that a continuous progression of wheat exists in the United States starting in at the end of August in the southwestern-most regions and moving progressively north until the last of the spring wheat is planted in April in the northern U.S.

Different types and cultivars vary in their response to different management practices and adaptability to different conditions. Our study only considered phenological timing. Whereas this timing is acknowledged to be variable, we used average heat units requirements within a sowing region to forecast the timing of the reproductive period (Feekes' scale periods 10.1–10.3)—the only wheat stage susceptible to karnal bunt.

Temperature and moisture are the two key factors that will lead to disease development in the presence of a susceptible host. In addition to the observation of weather data, the use of irrigation may increase relative humidity, especially when irrigation is applied at or near the reproductive or susceptible period.

Areas sown to wheat in the United States (as relevant to regulatory activities) are monitored by the Cooperative Agricultural Pest Survey (CAPS) survey program (D. McNeal, personal communication).

Determination of Risk Zones: A Simulation Model and GIS-Based Approach

All risk-relevant information was integrated into a system capable of providing categorizations of production areas relative to their susceptibility to karnal bunt. The first step was to determine the likelihood of the pathogen establishing successfully in different regions of the United States. Given what is known about *T. indica*, the following elements were critical for completing the missing elements of the risk assessment: determination of host-susceptible periods, determination of weather conditions, and determination of disease potential by region. A wheat degree-day model was used to predict wheat anthesis regions using weather and planting dates as input data. Historical weather data was thus summarized and isothermic regions were produced. The weather isotherms were constructed following temperature parameters of the disease index proposed by Jhorar and colleagues.[17] The final categorization of susceptibility regions was obtained by overlaying anthesis regions with predicted disease index maps.

Forecasting Phenology

The model of Rickman and colleagues,[18] MOD3WHEAT, was modified (R. Stefansky, personal communication) and used to forecast anthesis dates. Necessary input included planting dates and weather conditions. A basic step was the determination of wheat growing areas, types of wheat grown, and production practices. Planting dates were obtained from the National Wheat Grower's Association website. Weather data were obtained from public records [e.g., National Geophysical Data Center (NGDC) and the National Oceanic and Atmospheric Administration (NOAA)]. The model was run for each weather station. Weather records used in this analysis consisted of ten-year averages. The model ran on a daily time step based on these historical conditions.

Weather Records and Analysis

Current climatic trends show that long-term historical weather may not be representative of current conditions.[19] Coakley and McDaniel[20] have suggested a minimum of eight years of historical weather data be used in plant disease forecasting. In this study, United States weather records from 1986 to 1997 were summarized for 9,068 weather stations nationwide (these stations include locations in Alaska, Hawaii, and Puerto Rico, which were excluded in this risk assessment). The distribution of stations was more or less uniform with some two to four stations for each county. The density of stations tends to follow the density of counties. That is, where there are more counties, there are more reporting stations. Weather records were daily and the fields included temperature maximum, temperature minimum, precipitation, and snowfall. A second metafile including geographic reference information (latitude, longitude, and elevation) was also produced for all 9,068 stations. Calculating the daily averages for each station over all ten years produced a third file. Where ten years of data were not available for a given station, the available data were used and the average adjusted to reflect available data. This average file was used in the analysis and categorizations described. In order to obtain data for points in between weather stations, an interpolation algorithm was used. Interpolation is a process of estimating values that lie intermediate to known measurements. Interpolation yields new values that form a continuous surface between observed data. The value of the interpolated numbers depends on the data itself and on the assumptions that the observations capture the main factors responsible for the observed differences. Different types of spatial interpolation methods exist and these include inverse distance weighting, triangulation, and rectangular interpolation.[21] A triangulation interpolation process was used in this study.[22] Triangulation with smoothing was chosen because the values used in the interpolation can be inferred from maps by simple inspection and provides a quick method of verification. Inverse distance weighting was also tested in this study and provided the same categorizations as the triangulation method.

Resolution and Scale. Resolution refers to the density of observations used to characterize a region. The scale of a study is defined by the unit of observation/analysis and can be an individual plant, field, farm, county, landscape, etc. The total number of observations or "points" per area is also referred to as "the granularity" of the study. An important consideration was thus the resolution or level at which risk zones should be established. Historical weather data and wheat distribution may be examined at scales ranging from individual fields to multi-state wheat-growing "belts." However, the purpose of a risk assessment is the support of (regulatory) risk management/risk mitigation procedures. It thus behooves the production and trade industries to regulate the smallest area that will result in mitigated risk. The assessment could then be conducted using the smallest possible resolution. There exists another important reason to use as fine a granularity as possible: weather is highly variable. The characterization of a state, for example, with data from a single weather station would be informative but very limited in its usefulness. To the extent possible, the number of monitoring stations used in a wide-area forecast should be maximized.

Estimating Disease Index and Establishing Risk Zones

Data for *T. indica* development, specifically the production of secondary, grain-infecting spores, originate mostly from controlled chamber studies. The field model of Jhorar and colleagues[17] was used as a general guideline for two reasons: (1) it was the best field evidence we found on the behavior of this pathogen on wheat and (2) it has been adopted as part of the European Plant Protection Organization (EPPO) standard for karnal bunt. Jhorar and colleagues noted *T. indica* establishment only for the range of conditions bounded by 14–20°C and by an HTI of 2.2–3.3. Conditions for establishment of *T. indica* were not considered favorable outside those weather parameters.

The conservative approach adopted here (conservative in the sense of erring on the side of overpredicting establishment or, expressed differently, of predicting establishment where it may not occur if all conditions were considered) assumes that the relative humidity is always optimal. The boundaries constructed assuming relative humidity is always optimal were: low, maximum temperature greater than 20°C (upper bound for no disease) or less than 14°C (lower bound for no disease); medium, maximum temperature 14–16°C (lower bound for disease development) or 18–20°C (upper bound for disease development); high, maximum temperature 16–18°C (optimal conditions). In terms of probability, a subjective assessment where the probability is expressed as high when the conditions are optimal would correspond to a range of 0.95 to 0.999; probability is expressed as medium when conditions deviate from optimal and corresponds to a probability of 0.80 to 0.94; low corresponds to a probability of 0.0 to 0.79. Note that these estimations of probabilities are only provided here to illustrate the probabilistic correspondence of this phase to the general risk model. However, our analysis intends to issue risk-management decisions based on the qualitative categories of high, medium, and low. Thus, there were no model calculations performed with the expressed numeric probabilities.

Another reason for using this approach (i.e., assuming humidity is optimal) lies in the fact that relative humidity data were not available for most of the reporting stations. Using indirect estimates of relative humidity would not be expected to provide reliable estimators. This type of data is prone to incorrect predictions (R. Waight, personal communication).

A GIS environment was used to overlay the areas of forecast anthesis with the areas of predicted high, medium, and low disease risk. The joint occurrence of conditions ideal for *T. indica* with susceptible stages of the crop is the basis of the risk categorization.

Discussion of the Case Study

Results of Phenology Simulation

A file containing forecast anthesis dates by county was produced by the phenology model. Anthesis regions were then calculated by grouping all counties where wheat was in anthesis during the same two-week period, approximately.

Zones of Equal Colonization Probability

The predictions from the phenology model (i.e., what regions would be in anthesis and during which period) were overlaid with maps showing isotherm regions. The isotherm regions were produced based on the susceptibility regions described above and provide color-coded indications of susceptibility. The risk zones were calculated for different periods corresponding to anthesis periods for wheat sown in winter and spring. Isosusceptibility or isopathogenicity regions based on isotherms were categorized as described. Susceptible areas during this period (April 15–April 30) were limited to the northern portion of the United States.

The anthesis (isophenology) regions were overlaid on the isopathogenicity regions. Only the intersection of these regions (isophenology and isopathogenicity) is risk-relevant. Interestingly, small portions of Imperial and San Diego counties in California may be considered medium to high risk. Pima and Yavapai counties in Arizona also show reduced areas of high and medium risk. All other areas are considered low risk. Note from the wheat production map that no wheat is reported from Yavapai county.

The isophenology regions and isopathogenicity regions were investigated for the period May 1–May 14. It was noted that Newton county in Arkansas and the eastern edge of Tennessee and western edge of North Carolina may be considered in the medium risk category. All other areas are considered of low risk. Similarly, isophenology and isopathogenicity regions were estimated for the period May 15–May 30. The central region in Colorado falls into the medium and high risk (areas of Jackson, Grand, Larimer, and Park counties) categories. All other areas are considered low risk. Isophenology and isopathogenicity regions were also estimated for the period June 1–June 14. Only very small portions of Clackamass and Hood River counties in Oregon are in the medium-risk categories. All other areas are considered of low risk. For the period June 15–June 20, isophenology and isopathogenicity regions were estimated as above. Small portions of Idaho and Valley counties in Idaho are in the medium-risk categories, as are small portions of Glacier and Pondera counties in Montana and small parts of Park and Teton counties in Wyoming. All other areas are considered of low risk.

Isophenology and isopathogenicity regions were estimated for the period June 21–June 30 (spring-planted wheat). Small portions of Eagle county, Colorado and Sheridan, Wyoming can be considered of medium risk. All other areas are considered of low risk. Finally, isophenology and isopathogenicity regions were estimated for the period July 1–July 15 (spring-planted wheat). All areas were considered low risk.

Risk Regions and Risk Perspective

Risk as described in the general risk model is not simply the assessment of the probability of establishment (and other model elements) but also includes an assessment of the outcome of establishment. In the case of karnal bunt, whenever the disease is detected (as evidenced by the presence of bunted kernels), we can expect that this will trigger regulatory quarantines, eradication activities over an area larger than the area of detection, and the potential disruption of trade. This implies that the deleterious outcome of karnal bunt establishment will be very high until risk mitigation (management measures) procedures have been implemented.

For the purposes of establishing risk zones based on the probability of establishment regions as described above, we considered that if a given county grows sufficient wheat to warrant inclusion in the CAPS 1996–1997 survey, then it will be considered an economically significant region. Further, it is assumed that a highly significant and negative outcome will result should karnal bunt be detected.

As noted, the detection of any level of bunted kernels in a new county in the United States may trigger immediate regulatory action and will deleteriously affect wheat production and trade. Another detailed analysis of economic impact is trivial at this time given the known past trade and economic impact of this pest. Thus the regions described are not only "probability of establishment" zones, they are also "risk zones." Nevertheless, it is important to note the areas where the impact of the disease would be strongest. A key economically relevant factor is the amount and economic importance of the crop produced by regions (data not shown). These data may further refine resource allocation within the framework of risk mitigation activities.

We note that previous risk assessments for karnal bunt in the United States have differed in outcome from the present analysis. In a comprehensive but qualitative approach, Schall[23,24] concluded that the risk for karnal bunt was high over much broader regions than those identified in this study.

Colonization versus Spread

Spread or "dispersal of a disease is the movement of propagative units (inoculum) from the place where they are formed."[25] For karnal bunt, our assessment so far has assumed the potential presence of spores in the entire United States. This analysis as discussed above implies maximum spread potential because of the assumptions noted. Knowing the area of maximum spread is useful in forecasting the boundaries of the disease in the event it were to become maximally widespread and uncontrolled by risk mitigation procedures. Maximum spread potential provides the upper boundary of expected disease spread.

The assumption of widespread inoculum/disease is however, untrue as demonstrated by the results of the 1996–1997 surveys (D. McNeal and J. Pheasant, personal communication). The only known areas of propagule development are the areas reporting bunted kernels in Arizona and San Saba County, Texas.

A more precise forecast of potential spread would examine human transport from areas of known infection and potential wind-borne spread. Because of existing quarantines and imposed risk-mitigation procedures,[13] it is not expected that human activity (e.g., transport of wheat to mills and distribution of seed) will be a significant factor in disease spread. An analysis of wind-borne propagule dispersion will be conducted as an addendum to this study but will obviously not change the maximum spread potential regions as described by the risk zones established in this analysis.

Policy Relevance

The results of this assessment provide indications as to the expected patterns of karnal bunt infestation over the long term. Individual years may of course vary and the risk of any given year may deviate significantly from the forecasted trends. Given the poor predictability of weather, the use of historical patterns and expected trends is a scientifically sound approach.

The risk zones identified should be used in planning survey programs; for the allocation of resources by identified risk region and in delimiting the areas of interest from a regulatory perspective. The absence of conditions favorable to karnal bunt in most growing areas identified using the epidemiological and spatial analytic approach developed here should be presented as scientific evidence to support the results of negative surveys. The detail included in this analysis is beyond that which has traditionally been associated with APHIS-PPQ risk assessments. There is a reason for this: karnal bunt is a pest of export significance. Given the current global trade agreements, the burden of proof lies largely with the exporting country. The spatial resolution represented in this assessment may not be appropriate for pests and commodities being considered for import. In such cases, coarser-grained regions may be used until the interested party (e.g., an agribusiness desiring to export to the United States) provides data and procedures similar to those described herein in the spirit of reciprocity.

Summary for the Case Study

The present case analysis addressed the probability of colonization and spread for karnal bunt and integrated the results with other risk model elements (e.g., wheat production regions, yield by region, value of the crop) in order to produce "risk regions" for risk management.

The basic approach used to produce these risk regions was based on the epidemiological notion that in order for a disease to develop there must exist: (1) a susceptible host, (2) environmental conditions conducive for pest development, and (3) a virulent pathogen. Wheat susceptibility was determined by using a phenological model (a computer-based simulation model to predict when different plant physiological events, such as flowering, occur).

Environmental conditions were determined by analyzing weather patterns for ten years from 9,068 weather monitoring sites uniformly distributed over the entire continental United States. Pathogen inoculum was assumed uniformly present. All necessary data types were integrated using a GIS.

Where an intersection of anthesis and weather conditions conducive to the disease was noted, a region of significant probability of colonization (high, medium, and low) was determined. A categorization of the probability of colonization into high, medium, and low areas was determined based on how close the average climatic conditions were optimal for karnal bunt. A further mapping of these probability-of-colonization regions over productivity and other base maps provided a final assessment of risk.

The following conclusions are based on the analysis described above. A detailed assessment of the risk of karnal bunt to U.S.-grown wheat confirms that the great majority of production regions are not highly susceptible to the development of the disease. Very limited areas have been identified where risk may be considered as medium. The analysis of weather patterns prevailing during the production of winter and spring wheat shows that wheat planting patterns are such that anthesis (the susceptible period) does not coincide with climatic conditions favorable to the disease. The areas of highest probability of colonization did not coincide with high production or high value regions and therefore cannot be considered high risk. The majority of wheat-producing areas in the United States corresponds to the lowest category of risk for karnal bunt. This statement is true for both winter- and spring-planted wheat.

Our main conclusion is that given U.S. weather patterns and wheat growing conditions, it is expected that karnal bunt will remain at levels "below regulatory concern" for the greater majority of the wheat production regions.

IMPLICATIONS OF THE CASE STUDY TO BIOLOGICAL TERRORISM AND AGRICULTURAL SAFETY

Technology is not a substitute for comprehensive knowledge of epidemiological complexities of pest-host interactions. However, maximally informed decisions will benefit from integrated environments that consider the implications of all data available. In today's information-rich environment these data include historical and real-time weather records, satellite imagery, Internet-based databases, and mathematical simulation models.

The existence of new data sources raises the problem of information overload and loss of perspective as to the relative value of individual data elements. The use of simulation models to integrate biological knowledge, geographic information systems to integrate spatial information, and machine learning techniques to identify trends in large, multi-dimensional search spaces are all promising elements that need to be considered in conducting comprehensive risk assessments and in the strategic planning of tactics to counteract the possibility of acts of biological terrorism.

The case study carefully avoids invoking traditional elements of a risk assessment, such as the use of stochastic variables, evaluation of perceived risks for given factors, and the automation of the process using a risk analysis software tool. The objective was to show how more mechanistic (mechanistic is used here to refer to approaches that retain a close one-to-one relationship with real-world elements as opposed to abstracting processes using high level stochastic distributions) approaches can provide useful insights and powerful communication alternatives to traditional probabilistic expressions.

It is proposed here that similar approaches need to be implemented to assess the risk posed by exotic pests to our key crops and native species. Some of the steps have already been initiated. The Center for Plant Health Science and Technology (APHIS-PPQ) has begun the production and maintenance of "top ten" foreign pest threats. This simple listing needs to be expanded with more formal mechanistic assessments as exemplified by the case study here. Further, the pest lists should be dynamic and comprehensive and the ranking of all pests posing a significant threat should be provided to U.S. scientists. A conversation as to the ranking itself and the standardized, mechanistic assessments of risks should be engaged.

CURRENT ISSUES AND PERSPECTIVES

Specific problems within current APHIS-PPQ capabilities have been identified: (1) The ability of APHIS to conduct risk assessments and coordinate efforts (e.g., PPQ and VS) is hampered by lack of information in a usable form to conduct the necessary spatial analysis, forecasting economic analyses, etc. There is no specific plan or infrastructure to deal with bioterrorism at a national scale. (2) Our environmental compliance monitoring in cases of regulatory action (e.g., during emergency programs) is limited by the fact that we do not have a core of trained GIS and GPS specialists. In the past, emergency programs (for example) have relied on expensive external expertise or the ad hoc, voluntary participation of a very few agency experts. (3) Our ability to analyze and report exotic pest distribution is hampered by lack of spatial analysis and GIS capabilities. This results in a lot of collected data not being used in decision making. (4) Our spatial analysis efforts in the agency, and more generally the use of the data collected for decision making, are uncoordinated and inefficient. Many activities/approaches are either redundant or, in some cases, obsolete. (5) Our ability to change the way our agency does business in the direction of an "information-based," accountable agency is hampered by our ability to analyze the data we have and to collect the data we need.

APPROACHES TO ADDRESS PROBLEM

Option 1. Establishment of a Systems Analysis and Geographic Information Systems Group

The group will consist of systems analysis and GIS experts located at regional hubs and satellite locations. These individuals will provide leadership in systems analysis and GIS and will develop protocols for detection and emergency action programs. All systems analysis and GIS activities will be linked forming an Internet-based central headquarters. This centralized system will serve as a clearinghouse for basic data layers, models, and other information as needed by each of the national programs and port operations. The systems analysis and GIS group will take advantage of current networking possibilities to join USDA-APHIS expertise located in different geographic regions. In addition to providing leadership in bioterrorism and emergency programs, the group will implement GIS and systems analysis operations for emergency programs as needed. The positions will be located at the western and eastern regional hubs. They will be managed by the PPQ unit but have broad service and coordination responsibilities for both animal and plant threats. A key element for this group will be to establish interagency collaborative agreements and to develop data-sharing capabilities relative to the issue of biological terrorism and alien species surveys.

The implementation of areawide procedures during an emergency using modern tools is best exemplified with a sample plan. This plan is briefly outlined below.

Sample Implementation of Spatial Analysis, Geographic Information Systems, and Areawide Monitoring in Support of Emergency Programs

Emergency programs are peculiar in the need for extremely rapid development and delivery of information to support the program. This information typically includes the spatial distribution of the outbreak, spatial trend analysis and forecasting, monitoring of survey and control activities, planning of regulatory actions, and others. Given that emergency programs involve a similar number of phases and actions, it is possible to plan the implementation of systems and spatial analysis to assist in strategic and tactical management.

The steps in implementing a systems and GIS system analyses in emergency programs are associated with three main areas: survey, control, and regulatory actions. Nevertheless, they will typically use the same basic data layers. The following steps are guidelines and are patterned after those used in establishing the karnal bunt[26] and other GIS programs. The guidelines assume that the necessary hardware/software is already in place and that a "systems analysis and geographic information systems" officer is on location.

(1) Establish basic data layers. Typically these layers consist of district or county boundaries, township-range-section grids, hydrology, city streets (if cities are included in outbreak area), domestic and international airports and shipping lanes, and major highways and roads. These data typically exist either in APHIS databases or can be purchased from the GIS vendors.

(2) Establish program-specific data layers. These data layers usually involve the boundaries of the production areas, the lots where the commodity is produced, or the address of the location of the invading pest. There are several ways to establish these program-specific data layers. In many cases, they may already exist. For example, in the case of citrus canker in Miami, the digitized maps of all city lots already existed in a database created by the local utility company (Florida Power and Light). The existence of this database provided an immediate ability to map the locations of the "positive trees." A second example is karnal bunt. In that case the database for field locations had already been established (at USDA-APHIS-PPQ Phoenix Methods Lab) for the boll weevil eradication program and had been periodically updated. This existing database was critical in the successful establishment of survey and regulatory procedures for karnal bunt. Where appropriate databases do not exist, point locations may be geographically referenced using GPS units, and groves may be delineated with GPS units or from high altitude photography and satellite images. Other sources of data include U.S. Geological Survey (USGS) Landsat and Satellite Pour L'Observation de la Terre (SPOT) satellite images [Centre Internationale en Recherche Agronomique et Development (CIRAD)], NASA, and USDA Agricultural Research Service (ARS) Remote Sensing Labs.
(3) Establish geo-referenced databases for survey activities, which can be integrated with the GIS component. Trap or scouting locations should be recorded including a location record (e.g., section, township-range), which can be automatically geographically coded by the GIS. This will enable rapid visualization of trap distribution or scouted fields and permit straightforward monitoring, quality control, and error trapping.
(4) Establish geo-referenced databases for regulatory activities. These databases will be similar to those used by the survey team. They may also include the location of distribution points, marketing areas, processing plants, and other points of entry/exit.
(5) Establish geo-referenced databases for control. Again these databases will be similar to the previous two databases, but will also show regulated areas and ground and aerial treatments. Control databases and maps will be used to monitor all ground and aerial control activities.
(6) Record survey, control, and regulatory activities periodically and conduct appropriate spatial analysis. Relevant spatial analysis may include aerial correspondence (e.g., overlaying current "positive infestation sites" with treated areas) and trend and distribution analysis.
(7) Periodically consult with emergency program director (state) and deputy program director (federal) as to the ongoing observations. Communicate and periodically distribute results to all program officers.

The establishment of emergency response structures will have side benefits and applications. This is important, since during non-emergency conditions these activities will enhance the benefits of and provide readiness to the emergency response teams. Some of these expanded applications include the formalization of the concept of certified pest-free zones for export enhancement and the establishment of risk-based resource allocation procedures at ports.

Option 2. Establishment of a Project within Existing Structures

The establishment of the Center for Plant Health Science and Technology (CPHST) and the regionalization of several field offices into hubs provide an opportunity to centralize core GIS operations and services within such structures. Further, the existence of advanced GIS and systems analysis capabilities within the Center for Epidemiology and Animal Health (APHIS-VS-CEAH, Fort Collins) makes it possible to consider a joint position. This position would have primary responsibilities for GIS support in PPQ. As a member of CEAH, this coordinator would have access to the GIS modeling infrastructure (hardware and software) and use its distribution facilities to support PPQ activities. The example implementation plan described for Option 1 applies to Option 2.

Option 3. Do Nothing

As part of the ongoing operations (domestic, trade, emergency) several groups are implementing GIS operations. Coordination efforts are sporadic and do not have program support or follow recognized agency mandates. Slow progress is expected but improvements may occur over time if sufficient funds are dedicated to developments in the area of systems analysis and GIS by independent programs (emergency, region, domestic programs).

ROLE OF APHIS IN SAFEGUARDING AGAINST AND RESPONDING TO BIOTERRORISM

In order to respond to biological terrorism threatening animal and plant production, APHIS will consider development of a "bioterrorism rapid response" strategy based on the "rapid response" teams already in place for domestic and non-terrorism related emergency operations. This enhanced rapid response effort will consist of teams trained in the coordination and execution of activities associated with an outbreak scenario. Further, an infrastructure will be maintained that periodically surveys ports and inland regions for the presence of exotic pest threats. This survey and monitoring effort will include cooperators in federal, state, and academic institutions. Different options to implement such teams will be considered and some of those options were detailed here.

APHIS' rapid response strategy will evolve to expand its activities in coordination with other emergency management agencies. Recently, emergency management and hazard management procedures from sister agencies (e.g., EPA, Forest Service, and FEMA) have been incorporated into rapid response procedures. APHIS will also evolve its information superstructure to include extensive application of geographic information systems, remote sensing, and forecasting models. Finally, the threat of key pests ranked according to perceived risk will be assessed with mathematical models and "what-if" scenarios analyzed to determine impact and mitigation practices.

The success of GIS-based monitoring in Agricultural Quarantine Inspection (AQI), domestic and emergency programs depends on the establishment of strong communication links between state and federal agencies. Overall, APHIS will be responsive to biological terrorism using both existing and evolving structures. These structures will include bioterrorism rapid response teams, extensive monitoring, wide area monitoring using remote sensing, field scouts, geographic information systems, and finally, a locally and globally networked information superstructure.

In this document we described the existing emergency structures, we presented a recent example that exemplifies potential new directions, and we identified problems and proposed new approaches.

REFERENCES

1. SIM, T. 1998. Plant pest quarantines: their role and use. *In* Bunts and Smuts of Wheat: An international symposium. V.S. Malik & M.E. Mathre, Eds.: 368–382. North American Plant Protection Org. Ottawa.
2. U.S. DEPARTMENT OF AGRICULTURE. 1998. READEO Manual. Regional Emergency Animal Disease Eradication Organization. USDA-APHIS-Veterinary Services. Emergency Programs. Riverdale, MD.
3. U.S. DEPARTMENT OF AGRICULTURE. 1998. Emergency Programs Manual. USDA-APHIS-PPQ. Manuals Unit. Frederick, MA.
4. NATIONAL ACADEMY OF SCIENCES/NATIONAL RESEARCH COUNCIL. 1983. Risk assessment in the federal government: managing the process: 17–50. Committee on the Institutional Means for Assessments of Risks to Public Health. Commission on Life Sciences. National Research Council. National Academy of Sciences and Food and Drug Administration. Washington, D.C. ("The red book").
5. ENVIRONMENTAL PROTECTION AGENCY. 1992. Framework for ecological risk assessment. EPA/630/r-92/001. Risk Assessment Forum. Office of Research and Development. Washington, D.C.
6. ENVIRONMENTAL PROTECTION AGENCY. 1996. Proposed guidelines for ecological risk assessment. Fed. Reg. **61**(175): 47552–47631.
7. FERENC, N. 1997. The development of agro-ecosystem ecological risk assessment. USDA ORACBA News **2**(5): 1–3.
8. AHL, N. 1997. Director's corner. USDA ORACBA News **2**(5): 5–6.
9. ROBERTS, T., A. AHL & R. MCDOWELL. 1995. Risk assessment for foodborne microbial hazards. *In* Tracking Foodborne Pathogens from Farm to Table. USDA-ERS Miscel. Publ. No. 1532: 96–115.
10. U.N. FOOD AND AGRICULTURAL ORGANIZATION (FAO). 1996. Guidelines for pest risk analysis. Part 1. Import regulations. International Standards for Phytosanitary Measures. Secretariat of the International Plant Protection Convention. FAO. Rome.
11. U.N. FOOD AND AGRICULTURAL ORGANIZATION (FAO). 1998. Pest risk analysis: Supplementary standard for quarantine pests (draft). P1-Import regulations. International Standards for Phytosanitary Measures. Secretariat of the International Plant Protection Convention. FAO. Rome.
12. ORR, R.L., S. COHEN & R. GRIFFIN. 1993. Generic non-indigenous pest risk assessment process. The generic process. Planning and Risk Analysis Systems. Policy and Program Development. USDA-APHIS. Washington, D.C.
13. BONDE, M., G. PETERSON, N. SCHAAD & J. SMILANICK. 1997. Karnal bunt of wheat. Plant Disease **81**(12): 1370–1377.
14. SINGH, D., R. SINGH, V. RAO, S.S. KARWASRA & M. BENIWAL. 1996. Relation between weather parameters and karnal bunt (*Neovossia indica*) in wheat (*Triticum aestivum*). Indian J. Agric. Sci. **66**(9): 522–525.

15. SMILANICK, J.L., J.A. HOFFMANN & M.H. ROYER. 1985. Effect of temperature, pH, light, and dessication on teliospore germination of *Tilletia indica*. Phytopathology **75:** 1428–1431.
16. AUJLA, S.S., Y.R. SHARMA, K. CHAND & S.S. SAWNEY. 1997. Influence of weather factors on the incidence and epidemiology of karnal bunt disease of wheat in the Punjab. Indian J. Ecol. **4**(1)**:** 71–74.
17. JHORAR, O.P., H.S. MAVI, G.S. HAHI, S.S. MATHAUDA & G. SINGH. 1992. A biometeorological model for forecasting karnal bunt disease of wheat. Plant Dis. Res. **2:** 204–209.
18. RICKMAN, R.W., S.E. WALDMAN & B. KLEPPER. 1996. MODWht3. Agron. J. **88:** 176–185.
19. CASTI, J. 1991. What Scientists Can Know about the Future. Morrow Publ. New York.
20. COAKLEY, S.M. & L.R. MCDANIEL. 1988. Quantifying how climatic factors affect variation in plant disease severity: a general method using a new way to analyze meteorological data. Climate Change **12:** 57–75.
21. WATSON, D.F. 1982. Contouring: a guide to the analysis and display of spatial data. Elsevier Sci., Inc. New York.
22. AKIMA, H. 1978. A method for bivariate interpolation and smooth surface fitting for irregularly distributed data points. ACM Trans. Mathematical Software **4**(2)**:** 148–159.
23. SCHALL, R.A. 1988. Karnal bunt: the risk to the American wheat crop. USDA-APHIS-PPQ. Riverdale, MD.
24. SCHALL, R.A. 1991. Pest risk analysis on Karnal bunt. USDA-APHIS-PPQ. Riverdale, MD.
25. CAMPBELL & MADDEN. 1990. Introduction to Plant Disease Epidemiology. J. Wiley and Sons. New York.
26. NAVE, R. 1998. Overview of Karnal bunt quarantine. *In* Bunts and Smuts of Wheat. V.S. Malik & D. Mathre, Eds.: 397–404. North American Plant Protection Organization. Ottawa.

Guarding against Natural Threats and Terrorist Attacks
An Industry Perspective

D. STOLTE[a] AND K.E. OLSON[b,c]

[a]*American Farm Bureau Federation, 600 Maryland Avenue SW, Suite 800, Washington, D.C. 20024, USA*
[b]*American Farm Bureau Federation, Public Policy Division, 225 Touhy Avenue, Park Ridge, Illinois 60068, USA*

INTRODUCTION

The American Farm Bureau Federation (AFBF) is the nation's largest general farm organization, with affiliates in all 50 states plus Puerto Rico. Our members produce virtually every commodity produced in the United States. Their objective is to continue to produce the safest, most abundant, and most affordable food supply in the world. As an organization we work to assure that they can meet this objective. In order to accomplish this, they need an agricultural research system that provides cutting edge technology, a technology transfer system that makes the needed information available to them in a timely manner, a handling and distribution system that maintains the safety of the food, and a science-based inspection system to assure consumers of the safety of our food supply.

There is growing awareness of the potential threat of the use of biological weapons by terrorists. The 1995 sarin attack in the Tokyo subway system is the most widely known actual case, but others have occurred[1] and the potential exists for many more. In addition to potential terrorist attacks, agriculture and the public face an increased threat of the introduction of naturally occurring pests and diseases. Increased foreign travel and trade result in increased exposure to foreign animal disease, pathogens, and pests not found in the United States. In some instances it may be difficult to tell if an animal or plant disease outbreak was the result of planned terrorist activities or the result of inadequate biosecurity. We support increased international trade as a way to provide increased market opportunities for producers, but care must be taken to assure that it does not result in unacceptable risks to producers and the public.

CURRENT ISSUES

Of necessity, our members have and will continue to focus on the natural threats to our food supply. Assuring that we have the infrastructure needed to address "nat-

---

[c]Address correspondence to: Kenneth E. Olson, Public Policy Division, American Farm Bureau Federation, 225 Touhy Avenue, Park Ridge, Illinois 60068. Telephone: 847-689-8743; Fax: 847-685-8969.

ural" threats, will better position us to deal with any problems introduced by terrorists. Producers are addressing this issue in a variety of ways, but they recognize that they will need to rely on others to protect us from direct terrorist activities.

Major issues that involve naturally occurring threats that currently face U.S. agriculture are discussed below.

Food-Borne Pathogens

Food safety is an issue of increasing public awareness and concern. Media coverage brings frequent reports of outbreaks from around the nation and the world. Sources may be traced to either plant or animal products, so all producers need to be prepared to address the issue. While questions do exist over the actual number of cases and the origin of the causative agents, food-borne pathogens present an issue that must be addressed. We are working toward that end.

Antimicrobial Resistance of Pathogens

An issue of increasing concern is the development of antibiotic resistance in pathogens. This has implications for farm animals, but the greater public concern is that these organisms may be transferred from animals to humans. This concern was the focus for a recent review by the National Academy of Science.[2] A strain of *Salmonella typhimurium* DT104, which is resistant to five antibiotics and is being seen with increasing frequency, is the most commonly cited concern in this area.

Introduction of Foreign Animal Diseases

Recent trade agreements include provisions to expedite the movement of products between nations. We support these agreements because they offer the potential of new markets. However, we recognize that by opening our borders through the regionalization process that has been implemented by the USDA, producers face some increased risk of introduction of a foreign animal disease. The increasing level of international travel also increases the potential of an inadvertent introduction of a foreign animal disease.

Zoonotic Diseases of Animals

We are close to completing the eradication of two long-standing zoonotic diseases from the United States, brucellosis and bovine tuberculosis. Efforts must be made to assure that the eradication is complete, and that appropriate monitoring and surveillance programs are in place to prevent the reintroduction of these diseases. We also need to have the infrastructure in place to deal with other existing or emerging zoonotic diseases.

Potential Loss of Pesticides

Since 1988, the total number of crop protection tools available to farmers has decreased more than 50%, from about 40,000 in 1988 to less than 20,000 today. Although the need for conventional pesticides is reduced with the development of some genetically engineered crops, such as corn, soybeans, and cotton, conventional pesticides are still essential for production of all conventional crops, including fruits and vegetables. With biotech crops, conventional pesticide alternatives are also essential for resistance management in pest populations.

Conventional pesticides are also key building blocks of integrated pest management (IPM) programs. Through the use of IPM, farmers have dramatically increased their use of alternative pest controls such as crop rotation, natural predators, and biological controls. Although IPM greatly reduces the overall use of pesticides, it cannot exist without occasional chemical controls during extreme infestations or when other practices fail. Many of these key IPM building blocks are threatened by the Environmental Protection Agency's (EPA) implementation of the Food Quality Protection Act (FQPA).

The AFBF is very concerned that entire classes of crop protection products, particularly the organophosphates and carbamates, are threatened with cancellation or restriction by the EPA's implementation of the FQPA. Such action will severely disrupt U.S. farm operations and will reduce the effectiveness and use of alternative controls. From an international perspective, cancellation of some crop protection products could displace domestically produced agricultural commodities with imports from other countries with less stringent pesticide and food safety regulations. This could also leave our crops more vulnerable to potential terrorist attacks.

Limited Numbers of Approved Animal Health Products

The number of animal health products available for use with animals is limited and prospects for additional new products do not appear great. It is important that the access to and viability of existing products be maintained and that the provisions of the Animal Drug Availability Act be implemented to facilitate access to additional products. This is needed to allow producers to care for their animals appropriately. It is likely to also be important if there were a bioterrorist attack that affected animal agriculture.

AFBF ACTIVITIES

The issues that have been identified generally concern naturally occurring events, although it is quite possible that bioterrorists could work in very similar areas. If we are prepared to deal with these "natural" issues, we will also be much better prepared to deal with problems introduced by terrorists. To address these issues there are a number of areas where AFBF is actively involved.

Animal Health Emergency Management Planning

Changing trade laws and static budgets for USDA have eroded the infrastructure that has protected animal agriculture. The potential shortcomings of our existing system were identified in a 1996 survey of state veterinarians.[3] The findings have prompted work by USDA's Animal and Plant Health Inspection Service (APHIS) and animal agriculture organizations to develop jointly an improved system to deal with the issue.[4] It is designed to prevent the outbreak of diseases to the greatest extent possible. If an outbreak were to occur, it seeks to limit the scope of the outbreak.

In order to do this, coordination as well as adequate funding and personnel are needed at the national, state, and local levels. AFBF has been actively involved with this effort from the start and will continue to play a role as the system evolves over the next few years. This system would likely play a vital role in the case of a bioterrorist attack.

Food Quality Protection Act

Workable implementation of the FQPA is now AFBF's top priority. We are very concerned that the EPA's current implementation strategy spells disaster for U.S. farmers and ranchers. Properly implemented, we believe the new law can work well for everyone, including consumers and farmers.

For satisfactory implementation, EPA must make major procedural changes. Most important, regulatory decisions affecting the registration and use of crop protection tools must be based upon sound science or at least good information. Unfortunately, risk assessments for new requirements of the law relative to extra margins of safety for children, common mechanisms of toxicity, drinking water, and residential exposure are being made with incomplete data or no data at all. This methodological flaw is resulting in unrealistic, overly conservative "default assumptions" that greatly exaggerate actual risk and threaten the use of safe and essential crop protection tools. The AFBF is now engaged in an intense effort for science-based implementation of FQPA that can work for farmers and consumers.

Individual Species Quality Assurance Programs

All livestock species have developed quality assurance (QA) programs dealing with issues such as residue avoidance, biosecurity, and animal handling.[5] AFBF supports these efforts, encourages producers to participate in them, and, in several cases, has been actively involved in the development of these programs. Widespread adoption of these programs by producers would help to limit the impact of a bioterrorist attack on animal agriculture.

Research Priorities

AFBF routinely works with others in industry and at USDA to define research priorities to address these issues. We support a broad portfolio of research, including development and dissemination of new animal vaccines and diagnostics, as well as plant protection tools.

Support for an Improved Food Safety System

Consumers want and need assurance of the safety of their food supply, but this is a multifaceted issue. We support the use of Hazard Analysis Critical Control Point (HACCP) programs for food inspection and as the basis for producer QA programs. We also support the use of processing technologies that can reduce risk. A final step is consumer awareness of their role in food handling and preparation to assure safety. If all of these pieces are in place, the risk of terrorists directly impacting our food supply will be reduced.

CONCLUSION

Being prepared to deal with natural threats and terrorist attacks will require coordination on many fronts. Producers can and are seeking to address those natural threats where they have some level of control. Even in this area coordination is required among producers, processors, handlers, government agencies, and the food distribution system to do the job that needs to be done. As we look to dealing with threats from terrorists, we also need to rely on others, such as the departments of Defense and State, to minimize these risks.

REFERENCES

1. ANDERSON, J.H. 1998. Microbes and mass casualties: Defending America against bioterrorism. Backgrounder No. 1182. The Heritage Foundation. Washington, D.C.
2. NATIONAL RESEARCH COUNCIL AND INSTITUTE OF MEDICINE. 1998. The Use of Drugs in Food Animals: Benefits and risks. National Academy Press. Washington, D.C.
3. MCCAPES, R. 1996. Preparedness—To stamp out foreign animal disease outbreaks. *In* Proceedings 80th Annual Meeting: 68–76. Livestock Conservation Institute. Bowling Green, KY.
4. ANNELLI, J. 1997. A national model for emergency animal health response. *In* Proceedings 21st Annual Meeting: 57–68. Livestock Conservation Institute. Bowling Green, KY.
5. UNITED STATES DEPARTMENT OF AGRICULTURE AND FOOD SAFETY INSPECTION SERVICE. 1997. Animal production food safety: An overview for FSIS employees. United States Department of Agriculture and Food Safety Inspection Service. Washington, D.C.

The Role of National Animal Health Emergency Planning

JOHN B. ADAMS[a]

Animal Health and Farm Services, National Milk Producers Federation, Arlington, Virginia 22201, USA

INTRODUCTION

The protection of animal health is an essential and basic element for ensuring the safety of the food supply of the United States. Protecting livestock and poultry from bioterrorism must become an essential element of strategic planning as this nation seeks to improve its defense capabilities.

While the threat of bioterrorism against animal agriculture may not be a new threat, the strategic importance of animal livestock to the nation's welfare and economy has never been more important. As a $100+ billion industry, American animal agriculture is and will remain a most important nutritional food source for the American population and for growing export markets. Billions of dollars of animal by-products account for ingredients in many other essential foods and products that help sustain our economy and military, not to mention the millions of jobs created throughout the entire food distribution chain.

As we increase the numbers of animals at fewer production facilities to provide the abundance and consistent quality demanded by consumers, concentration makes the industry more vulnerable to a terrorist attack. On the other hand, as we design and introduce more prevention-based food safety systems into the food chain, we increase the opportunity to prevent bioterrorist attacks.

The development of a viable National Animal Health Emergency Management System for the United States has become, in recent years, a shared partnership between the U.S. Department of Agriculture's Animal Plant Health Inspection Service (APHIS), the American Veterinary Medical Association, U.S. Animal Health Association (USAHA), and the animal industry as represented by the Animal Agriculture Coalition (AAC). This paper addresses this process and provides some perspective on the importance of emergency animal health management for protecting American food security.

THE DEVELOPMENT OF A NATIONAL ANIMAL HEALTH EMERGENCY MANAGEMENT SYSTEM

Major progress has been made over the past three years to develop a national strategy for animal health emergency management planning. The proposed strategic plan

---

[a]Address correspondence to: John B. Adams, Director of Animal Health And Farm Services, National Milk Producers Federation, 2101 Wilson Blvd. Suite 400, Arlington, VA 22201; Telephone: (703)484-3623; Fax: (703) 841-9328.

incorporates four basic elements utilized by the Federal Emergency Management Agency (FEMA): planning, preparedness, response, and recovery. This four-part concept expands on the more traditional approach utilized by APHIS/USDA to "respond" to the threat of a foreign animal disease introduction. The new model relies on the partnership previously mentioned and defines emergency from a broader perspective: "any sudden negative impact by the appearance of a disease or pest or the perceived health risks poised by such an event to the public or to foreign trading partners." The reference to "any sudden negative impact" would certainly apply to a terrorist attack or any emerging or endemic disease event(s). Finally, the new model recognizes that animal health emergency management must be integrated at local, regional, and national levels to facilitate an efficient mode of operations. This has become critical for the USDA/APHIS budget is no longer adequate to address this essential need without the partnership being developed. If one considers just the response capability of APHIS, the Regional Emergency Animal Disease Eradication Organization (READO) group is now reduced to two geographic regions—Eastern and Western. In 1993 four READO teams existed. Before that as many as six teams existed.

If the new strategic plan is to succeed, clear goals and directives for each of the partners must be identified and appropriate leadership and training must be delivered at all levels to achieve the desired coordination. To address leadership, training, and coordination issues, the Steering Committee on National Animal Health Emergency Management has undertaken the development of minimum standards or benchmarks that can help guide the states in performing their constitutionally required effective first response. The subsequent uncertainties associated with the transfer of authority from state to federal organizations in the event of an emergency has been identified as "the largest area of vulnerability in the system."[1]

A major assessment process was begun in 1998 to determine the status of state prevention, preparedness, response, and recovery capabilities. From this assessment process, the Steering Committee plans to develop a set of minimal standards of performance for all partners or stakeholders in the system. Private veterinary practitioners, producer groups, and other partners will be able to better identify their particular roles and responsibilities once the assessment process is more complete. Setting minimum benchmarks or standards can help identify legal and regulatory roadblocks, necessary levels of training, and resource needs. Commensurate with the development of the assessment process, the Steering Committee understands the need to begin to develop strong working partnerships and networks involving different government agencies, producers, private practitioners, diagnostic laboratories, industry-affiliated groups, and the public. To accommodate the planning for this wider and essential linkage between many involved or concerned groups (including the military, FEMA, and the Department of Health and Human Services (HHS)), the Steering Committee is proposing the formation of a National Council on Animal Health Emergency Management. Beyond the major goal of overall planning and coordination, the Council would focus on potential emerging national and international animal health emergencies, including bioterrorism, and assure the appropriate levels of preparedness, response, and recovery capabilities under the direction of a National Animal Health Emergency Strategic Plan. The partners and stakeholder members of the Council would be responsible for keeping the strategic plan updated and assuring its implementation.

Unless an adequate, functional, well-coordinated national animal health emergency management plan and an infrastructure are quickly developed and implemented in the United States, this country will not be adequately prepared for the future reality and consequences of increased trade, animal movements, greater risks of foreign animal disease introduction, and the increased risk of bioterrorism. Critical to the success of this overall national animal health emergency management planning initiative will be the strong endorsement by all stakeholders, including producers, veterinarians, allied industry groups, and both state and federal governments. The recent debacle in Great Britain with regard to the lack of appropriate management of the outbreak of bovine spongiform encephalopathy should serve as a wake-up call for the United States to develop and quickly implement a coordinated national animal health emergency strategic plan. Such a system will assure that strategic planning will occur to prevent, to the extent possible, economic or environmental impacts that could adversely disrupt our food production and distribution systems. Such a National Animal Health Emergency Management System can integrate and embody herd and flock biosecurity that is being implemented by many producers. It can also reinforce the development and implementation of Hazard Analysis Critical Control Point (HACCP) food safety systems being employed in more and more of our food processing plants.

SUMMARY

Bioterrorism is one of many contingencies that we must be better prepared to prevent from adversely impacting our food security system. Education and awareness, prevention-based training, increased monitoring and surveillance, and greater focus on coordination between government agencies at both the federal and state levels will be necessary to develop a more prevention-based national food security system. The formation of the Steering Committee on National Animal Health Emergency Management is the beginning of a profound new process. The animal industry is driving the formation of a new national animal health emergency management system. It is designed to coordinate the emergency prevention and response activities of all partners, including government (state and federal), livestock producers, and veterinary practitioners. The goal is to develop a national coordinated strategy that places primary emphasis on preventing a broader spectrum of hazards from occurring throughout the food production chain. Coordination and integration of efforts are necessary with many different stakeholders, including government agencies other than U.S. Department of Agriculture. Through a proposed National Animal Health Emergency Management Council, all stakeholders can assist in the development of a national strategic plan that can better protect our food security system.

REFERENCES

1. 1997. Quadrilateral Review of the United States Animal Health Emergency Management System. This report was developed for USDA's Secretary's Advisory Committee for Foreign Animal and Poultry Diseases.

Industry Concerns and Partnerships to Address Emerging Issues

BETH LAUTNER

Science and Technology, National Pork Producers Council, P.O. Box 10383, Des Moines, Iowa 50306, USA

INTRODUCTION

In the past, government officials were seen as solely responsible for animal health programs that would address emerging issues such as foreign animal diseases. Recently, animal agriculture has recognized the need to work in partnership with the government to address these types of issues, which have significant consequences for the animal industries.

The occurrence of a foreign animal disease such as classical swine fever, African swine fever, or foot-and-mouth disease in U.S. livestock herds would be devastating to animal agriculture. U.S. producers and state and federal animal health officials must not become complacent about the potential risks to U.S. animals. It is critical that all precautions are taken by the industry and government to protect domestic herds from the incursion of a foreign animal disease. Just as individual farms establish biosecurity guidelines to keep certain domestic diseases from entering their herd from other herds, the U.S. must have a comprehensive nationwide biosecurity system to prevent the introduction of foreign animal diseases from other countries. Both government and industry have a role in the implementation of this biosecurity system.

RISK OF A FOREIGN ANIMAL DISEASE

Recent events, such as the classical swine fever outbreaks in the Netherlands, Haiti, and the Dominican Republic and foot-and mouth disease outbreaks in Taiwan have heightened interest in what the U.S. is doing to prevent the incursion of a foreign animal disease. Many would view the risk of entry of a foreign animal disease as increasing.

The increased risk is attributed to several factors. There are many international visitors to the U.S. that may have contact with U.S. herds and recently have been on farms in their own country. Tour groups of producers from other countries to view U.S. animal agriculture are extremely common. In addition, many U.S. producers travel internationally to view production practices in other countries. Veterinarians and veterinary students are frequently consulting with or conducting research in other countries and then returning to work with U.S. herds. With "free trade" agreements, there is the potential for an increase in the flow of various products from an increased number of countries. This increase in diversity may contribute to increased risk. A major concern is the illegal importation of products that may carry infectious diseases. Border controls are critical to intercept this type of contraband. Animals may also be illegally imported and therefore not be subjected to U.S. import testing requirements. Recently, there has been a heightened awareness of the potential for bioterrorism to be directed as animal populations.

OVERVIEW OF THE U.S. PORK INDUSTRY

When looking at the potential impact of a foreign animal disease on a livestock industry, it is necessary to understand the structure of the industry and current production practices. The trend in the pork industry is to have fewer farms and larger numbers of animals on each farm. In addition, to optimize health, multi-site production is becoming more common in the industry. This involves having one site for the breeding herd with separate locations for both the nursery and growing pigs. These locations, while under the same management, may be located in different states. Therefore, one main breeding herd may move groups of pigs to a different state or several states for a stage of their life cycle. This would allow a foreign animal disease to potentially be moved rapidly to different geographic locations. In addition, the number of sources of genetic material (live animals or semen) is becoming more limited. This could facilitate the rapid spread of a foreign animal disease through the industry.

It is important for officials to understand the structure of the animal industries to ensure that unique characteristics are taken into account when strategic plans are being developed to address the incursion of a foreign animal disease.

ECONOMIC IMPACT OF FOREIGN ANIMAL DISEASES

If a foreign animal disease became established in U.S. herds, the economic consequences to producers and the public would be severe. Agricultural-related businesses, such as the feed industry, livestock markets, packers, pharmaceutical companies, and private practitioners, would also experience business losses.

One of the severe economic consequences of an incursion of a foreign animal disease into the U.S. is the immediate halt to exports. U.S. livestock industries are becoming increasingly dependent on export markets. The U.S. pork industry currently exports over 6% of its annual production with a value of more than $1.0 billion. The U.S. is the world's second largest exporter of pork. The cessation of exports due to an embargo or animal disease outbreak in 1997 would have caused cash hog prices to plummet by $6.29 per hundredweight, or more than $15 per market hog.

Currently, U.S. consumers spend only about $0.12 of every $1 of income on food. This is due in part to the efficient production of animal protein. A foreign animal disease could substantially disrupt the food supply in this country and result in increased consumer prices. For example, while the 1983–1984 eradication of highly pathogenic avian influenza in the U.S. resulted in the destruction of more than 17 million birds and cost $86 (1996$) million, the consumers' cost of poultry and eggs increased $548 million (1996$) due to this outbreak. More than $7 million was lost in wages (1996$) in related businesses.

The 1951–1952 foot-and-mouth disease outbreak in Canada involved only about 2,000 animals and incurred $2 million (1987$) in eradication costs. Because of international embargoes, the economic impact to producers was about $2 billion (1987$).

Animal agriculture groups are continuing to inform Congress and the public that investing in foreign animal disease prevention is not just a producer issue, it is in the best interests of protecting the efficient, inexpensive food supply currently available to American consumers.

EFFORTS TO MINIMIZE THE POTENTIAL FOR A FOREIGN ANIMAL DISEASE OUTBREAK

In recent years, animal industry groups, state animal health officials, and federal officials have been concerned about the nation's ability to prepare for and/or respond to emergency disease situations. Because of this concern, the Animal Agriculture Coalition (AAC) in early 1996 facilitated the formation of a Working Group on National Animal Health Emergency Management to discuss development of a new model for animal health emergency management in the U.S. Working Group members included the animal agriculture industry associations, Animal and Plant Health Inspection Service, state veterinarians, and the American Veterinary Medical Association. This model has been discussed earlier in this volume.

The AAC has also formed another working group. This group, the Working Group on Research and Diagnostic Needs for Animal Health Emergency Preparedness and Response, was organized in 1997. The purpose of the group is to identify the research and diagnostic needs for a new world-class national animal health emergency management program for the next century. Participants include the AAC, the U.S. Animal Health Association, the Agricultural Research Service, the Animal and Plant Health Inspection Service, and the Cooperative State Research, Education, and Extension Service.

The Working Group is reviewing current animal health research and diagnostic programs in emerging diseases, foreign animal diseases, and selected domestic diseases. The next step will be to determine future animal health research and diagnostic needs and develop strategies to address these needs with regard to funding, facilities, and staffing. The Working Group has toured several research facilities including the National Centre for Foreign Animal Disease in Winnipeg, Manitoba, the National Animal Disease Center in Ames, Iowa, and the National Veterinary Services Laboratories in Ames, Iowa and Plum Island, New York. The industry believes it is critical that adequate resources are provided for research and diagnostic needs for emerging and foreign animal diseases.

On a species-specific basis, another partnership between industry and government that will provide additional ways to address emerging diseases is the Swine Futures Project. The purpose of this project is to determine how Veterinary Services (VS) swine-related activities should be modified over time to meet the needs of the changing pork industry. The four-person team (two representatives from industry and two from VS) have identified one of the key areas for the project as the development of an effective emerging disease detection and response system. It has been noted that the changing industry structure could potentially allow an emerging animal disease to move rapidly into a large number of herds before detection. The project report outlines specific recommendations that would create a system to allow early detection of an emerging disease and a coordinated response if one were to occur.

CONCLUSION

The introduction of a foreign animal disease in the United States would have a very significant impact on American producers, businesses, and consumers. Every effort must be made to minimize the potential for an incursion of a foreign animal disease. American animal agriculture must continue to work proactively in partnerships with the U.S. Department of Agriculture and other federal agencies in these efforts.

International Economic Considerations Concerning Agricultural Diseases and Human Health Costs of Zoonotic Diseases

ALFONSO TORRES[a]

Plum Island Animal Disease Center, USDA, Agricultural Research Service, Greenport, New York 11944, USA

Throughout civilization humans have depended on animals for food, clothing, tools, traction, transportation, warfare, financial security, companionship, and even for sheer enjoyment. That dependency has not changed for thousands of years since humans learned to domesticate animals for their own survival. The difference today is that most people in industrialized urban areas do not realize or appreciate that dependency.

Animal diseases can negatively affect the number and availability of animals, their productivity, or their appearance.[1] A few centuries ago, animal diseases affected mostly individual owners or herdsmen, but did not have serious consequences on the larger community. A similar event today will not only have a negative impact on the animal owners, but more importantly, will significantly affect the general economy of the region, the entire nation, even a group of nations. The importance of animal diseases as an element affecting international trade of animals and animal products has reached its full impact level with the recent designation by the World Trade Organization (WTO) of the International Office of Epizootics (OIE) as the international agency in charge of establishing animal health standards upon which international commerce can institute restrictions to prevent the spread of animal diseases from one nation to another. It is important to point out that while the spread of human diseases around the world is due to the unrestricted movement of people across political boundaries, animal diseases are, for the most part, restricted to defined geographic areas of the world due to the implementation of animal importation requirements, quarantines, animal movement regulations, and by disease control measures that include mass vaccination campaigns and animal depopulation practices. A number of animal diseases have been eradicated from countries or even from continents around the world by aggressive, well-coordinated, long-term animal health campaigns. This is in contrast to the relatively few human diseases successfully eradicated from large areas of the world.

Other papers in this volume provide specific examples of current animal disease events that have caused serious economic impacts and disruptions in international trade of animals and animal products, as well as approaches taken to establish rela-

---

[a]Address correspondence to: Alfonso Torres, D.V.M., M.S., Ph.D., Deputy Administrator, USDA, Animal Plant and Health Inspection Service, Veterinary Services, Jamie L. Whitten Federal Building, Room 320-E, 12th and 14th Streets at Independence Avenue, SW, Washington, D.C. 20250; Telephone: 202-720-5193.
e-mail: Alfonso.Torres@usda.gov

tive regional risks to allow safe commerce of animals and animal products. The economic effect of zoonotic diseases affecting humans has also been discussed elsewhere in this volume. This paper uses examples of the past to illustrate how a highly contagious disease of ruminants has affected international economies, history, and the ecology of a whole continent.

The disease is rinderpest, or also known in many languages as "cattle plague." Rinderpest (RP) is a disease with morbidity and mortality rates for cattle and water buffalo of more than 90%. The course of RP from infection to death is less than two weeks. RP has been known as one of the most serious animal plagues since oral and written records have been maintained. Rinderpest swept westward through Europe out of Asia with the many waves of military campaigns between these two continents. Some of the earliest descriptions of the disease are found in writings describing the military campaigns of Charlemagne in the ninth century.[2] In those years, Asian oxen used for transport of military supplies acted as asymptomatic carriers of the RP virus and spread the disease to naive European cattle, causing extremely high mortality. Perhaps this was one of the first unintentional uses of an animal disease as a biological weapon. Between 1711 and 1769 more than 200 million head of cattle died of RP in western Europe alone. This is almost ten times the estimated mortality of humans during the Black Plague pandemics of the fourteenth and fifteenth centuries. The serious consequences of RP led, in large part, to the development of strategies and legislation dealing with the control of epizootics. In the early 1700s the College of Cardinals in Rome commissioned Giovanni Lancisi, the personal physician of Pope Clement XI, to provide advice on how to deal with RP. His recommendations were promulgated in 1714 and included the implementation of quarantines to arrest animal movements, the immediate slaughter and deep burial of infected animals. These actions were reinforced by another physician, Thomas Bates, Surgeon of His Majesty's Household in London, with the implementation of the concept of compensation for owners of affected cattle. The dawn of the industrial revolution and the use of steam power for the transport and commerce of cattle across long distances in Europe led to a second massive outbreak of RP that killed most cattle in Europe from 1857 to 1866. The devastating economic impact of these outbreaks demanded scientific action. The epizootics of the eighteenth and nineteenth centuries led to the establishment of the first veterinary schools in Europe, as well as to the first veterinary departments in several European countries. RP also gave rise to the concept of mass vaccination and to the development of the clinical thermometer for use as a tool for the early field diagnosis of this highly febrile disease.[2]

Although there may have been introductions of RP into the African continent before the late nineteenth century, it was the "Great Rinderpest Pandemic" of 1889–1897 that had severe impact on that continent. It is thought that RP was introduced to Africa by the Italian army in Eritrea sometime during 1887–1893. From there it eventually spread throughout Africa, resulting in massive mortality of domestic cattle, water buffalo, and many species of wildlife including African buffalo, giraffe, eland, oryx, and kudu. Mortality rates in cattle and wildlife game reached 90% in some parts of Africa. This pandemic had long-lasting continental consequences.[2,3] Dominant cattle-keeping tribes like the Masai in East Africa were weakened so much that they were eventually replaced politically and economically by agricultural tribes. The pandemic also permanently altered the ecological balance of game species in the large plains of East Africa.[3] RP was eventually controlled within most southern African countries but persisted within the nomadic subSaharan tribes. From there it emerged from time to time, especially during the first and the second world wars. A second pandemic in Africa occurred from 1969 to 1973. This was controlled after massive international aid launched a highly successful vaccination campaign that almost eliminated PR from Africa. Unfortunately, subsequent complacency and lack of coordination led in 1981–1984 to the reemergence of RP in many African countries. The serious reemergence of RP in Africa resulted in the creation of the Pan African Rinderpest Campaign (PARC) with major support from the European community.[3] PARC has successfully brought the disease under control through a renewed massive vaccination campaign, leaving only limited areas in Sudan, Ethiopia, and Kenya still affected by this disease.

In 1920, the OIE was created as an international agency charged with the responsibility of keeping track of the movements of RP and other serious animal diseases around the world. RP has continued to damage the cattle population every time that there are military confrontations, even as recently as during the Vietnam War and the contemporary conflicts in Middle East countries and in the Horn of Africa. RP still occurs in many other countries of the Middle East and the Indian subcontinent, particularly Pakistan. Thanks to the PARC and the Global Rinderpest Eradication Programme (GREP)[4] it is expected that RP will be the first major animal disease completely eliminated from the world by the year 2010. However, political and military conflicts in the original historical reservoirs of RP, today occupied by Pakistan, parts of India, Afghanistan, Russia, and other southern former Soviet Republics, pose a significant threat to the economic stability of intensive cattle producing countries of Eastern Europe.

It is hoped that this example has illustrated the destructive power that many animal diseases have and the devastating consequences if such diseases, still existing in many parts of the world, were to be introduced, deliberately or accidentally, into high-density animal production enterprises in our country.

REFERENCES

1. HANSON, R.P. & M.G. HANSON. 1983. Animal Disease Control. Chapter 2. The Iowa State University Press. Ames, IA.
2. GIBBS, E.P.J., ED. 1981. Virus Diseases of Food Animals. Volume II. Chapter 18. Academic Press Inc. London.
3. COETZER, J.A.W., G.R. THOMSON & R.C. TUSTIN, EDS. 1994. Infectious Diseases of Livestock. Chapter 74. Oxford University Press. Oxford.
4. U.N. FOOD AND AGRICULTURE ORGANIZATION. 1996. The World Without Rinderpest. FAO Animal Production and Health Paper: 129. Rome.

The Cost of Disease Eradication

Smallpox and Bovine Tuberculosis

ANN MARIE NELSON[a]

Division of AIDS Pathology and Emerging Infectious Disease, Department of Infectious and Parasitic Disease Pathology, Armed Forces Institute of Pathology, Washington, D.C. 20306-6000, USA

> ABSTRACT: Although eradication is the ideal approach to reduce the economic and human health costs of disease, there may be both short- and long-term consequences. A $300 million effort succeeded in completely eradicating smallpox in less than ten years. The campaign was effective because variola virus produced acute illness, had no carrier stage or non-human reservoirs, and had an effective vaccine that was used in combination with international surveillance and public education. Bovine tuberculosis was completely eradicated in many U.S. herds at a cost of $450 million over 50 years using a "test and slaughter" program combined with meat inspection. *Mycobacterium bovis* often does not produce acute disease, persists in the carrier stage, has multiple non-human reservoirs, and easily crosses species. No effective vaccine or centralized global surveillance or eradication programs currently exist. Control measures result in significant economic losses. Smallpox eradication had limited economic consequences but has left much of world's population highly susceptible to zoonotic orthopoxviruses and to the use of smallpox as a biologic weapon. The primary threat of *M. bovis* exists in wildlife that share watering holes or pasture land with domestic stock. In the developed world, surveillance can minimize risks, but one-third of the world's population lacks effective agricultural and food safety programs, leaving them at substantial risk for zoonotic infection by *M. bovis*.

INTRODUCTION

A significant number of zoonotic infections are directly or indirectly related to agricultural practices. Factors that influence the frequency and pattern of zoonoses in agrarian societies include sharing living spaces with livestock, movement of animal populations, limited medical and veterinary facilities, environmental disasters, and civil unrest. In industrial societies, zoonotic diseases are more often related to intensive livestock production, centralized food processing and distribution, recreational activities, pets (including exotic animals), and suburban intrusion into animal habitats. Zoonoses can be predominantly animal–animal cycles with incidental animal–human transmission (rabies), human–human cycles with incidental human-to-animal transmission (human tuberculosis), or animal–animal, human–human,

---

[a]Address correspondence to: Ann Marie Nelson, M.D., Chief, Division of AIDS Pathology and Emerging Infectious Diseases, Armed Forces Institute of Pathology, AFIP-CPS-B, 14th and Alaska, Room 4019, Washington, D.C. 20306-6000; Telephone: 202-782-2260; Fax: 202-782-9160.
 e-mail: NELSONA@afip.osd.mil

animal–human cycles (influenza, leprosy). Zoonotic infections may affect a specific host with accidental transmission to humans (monkeypox) or may easily cross species (bovine tuberculosis).[1] Any of these factors can and do alter our ability to control transmission.

Both the impact of zoonotic diseases and the efforts to control them are costly. Control strategies should consider the short- and long-term costs of the various options (from doing nothing to eradication). Eradication programs often have the highest short-term costs; failure to act may lead to enormous long-term costs. During this century we have attempted to eradicate childhood diseases through vaccination programs, waterborne infections by filtration and chlorination, mosquitoes with DDT, and air pollution by regulating emissions. Two of the largest and most costly efforts were the eradication campaigns for smallpox and bovine tuberculosis. There are significant differences in the host specificity, transmission, natural history, control measures, and socioeconomic consequences of these two diseases. These differences affected the strategies and the outcomes of the programs. Smallpox was strictly a human disease; this host specificity allowed for effective eradication. Eradication, rather than the disease itself, resulted in the emergence of zoonotic orthopoxvirus disease. Bovine tuberculosis, on the other hand, occurs predominantly in animals. Its ability to cross species makes it an important human and agricultural threat and limits the eradication effort.

SMALLPOX

Smallpox has been known for more than 3,000 years; it is an acute, highly contagious exanthema, which had worldwide distribution before eradication. The disease is limited to humans and has two clinical forms, variola major (highly virulent with 25% mortality) and variola minor or alastrim (mild with <1% mortality).[2] The disease affected people of all ages and both sexes; attack rates were higher in children. During the age of exploration, the introduction of smallpox into non-immune populations (Native Americans, for example) resulted in extremely high rates of mortality.

By the mid-1700s, European and Revolutionary War physicians inoculated material from smallpox pustules into uninfected individuals. Although "variolation" caused a mild form of smallpox, it prevented significant morbidity and mortality. Edward Jenner (1749–1823) discovered that cutaneous injection of cowpox would prevent smallpox by conferring passive immunity. Jenner coined the term vaccination, from the Latin for cow (*vacca*). Smallpox vaccination was initially used to curtail disease outbreaks, but was routinely given to children in most developing countries by the twentieth century.[3,4]

In 1967, the United Nations World Health Organization (WHO) launched a worldwide vaccination campaign against smallpox; at that time, 10 to 15 million cases of the disease occurred each year, with more than 2 million deaths. The last case of endemic smallpox occurred in Somalia in 1977. In 1979, after two years without a reported case of smallpox, WHO marked the disappearance of smallpox from the earth and recommended that countries stop vaccinating against the disease. WHO also requested the destruction of all viral stocks.[5,6]

Smallpox eradication was effective because variola virus (1) produced a distinctive, easily recognizable acute illness; (2) had no carrier stage; (3) had no non-human reservoir of endemic disease; and because (4) a stable, inexpensive vaccine was produced that conveyed long-term immunity; (5) large-scale vaccination campaigns and intensive case reporting and surveillance were conducted in more than 80 countries; (6) public education was actively promoted in all of these countries to overcome the hiding of cases and to break the cycle of smallpox transmission. Other factors that contributed to the success of this campaign include the fact that universal standards were imposed worldwide, there were no significant economic losses as a result of the campaign (other than direct costs), and both affected and non-affected countries committed to and participated in the goal of eradication.[5]

The total cost of smallpox eradication was $315 million between 1967–1980; $30 million was from the United States and the rest from developing countries. Efforts were centrally coordinated by WHO and were global in scope. Nearly 700 workers from 73 countries worked for WHO—150 at any given time—for periods of three to six months. They helped to coordinate the efforts of more than 200,000 local workers in those countries most affected.[5,6] The long-term benefits are great in public health as well as economic sectors. Since the eradication of smallpox in 1977, the U.S. Public Health Service estimates a monthly savings of more than $30 million in material and human resources (cost of production, administration of the vaccine, treatment of adverse reactions, record-keeping and surveillance, immigration control, quarantine, and education) (Joel Breman, personal communication). The long-term consequences of eradication are a direct outcome of its success. Much of the world's population is unvaccinated and therefore at risk for zoonotic orthopoxviruses.

In addition to smallpox, the orthopoxviruses include buffalopox (India, Egypt, Indonesia), camelpox (Africa, Asia), cowpox (Europe, western Asia), monkeypox (West and Central Africa, lab monkeys), and vaccinia (worldwide).[3] Cowpox, camelpox, and buffalopox can cause contact lesions in animal handlers, but are usually not considered human diseases.

The most important orthopoxvirus causing human disease is monkeypox, a condition so named because it was discovered in laboratory primates in 1958 in Copenhagen.[3] The first cases of human monkeypox were identified in the Democratic Republic of the Congo (formerly Zaire) in 1970. By 1979, 55 cases were reported in central and west Africa.[7,8] The disease has a clinical appearance nearly indistinguishable from smallpox, but seems to have less morbidity and mortality. With the eradication of smallpox and the cessation of vaccination, the WHO was concerned about the potential for human disease. From 1980 to 1986, monkeypox was intensely studied in the Congo. An additional 346 cases were identified and the fatality rate was 10% (15% in non-vaccinated children <5 years old). Seventy-two percent were primary cases; case workers reported only five generations of spread. The 9% attack rate in non-vaccinated household contacts was significantly less than the 25–40% rate for smallpox. Although monkeys often test seropositive and may transmit disease to humans, they are not the natural reservoir.[7,8] The major reservoirs are several species of forest squirrels. Hunting and deforestation for agriculture play important roles in transmission. (Delfi Messinger, personal communication). WHO is currently investigating a new outbreak. Between 1996 and 1998, a possible 500 new cases were reported in the Kasai region of Congo. Serologic studies revealed that many of these cases were actually varicella. Animal studies now indicate that antiviral drugs are effective in treating the infection. Because of this and the apparent low human-to-human transmission, monkeypox is not considered a significant natural or terrorist threat at this time.[7]

The most potentially devastating consequence of smallpox eradication is the fact that much of the world's population in now susceptible to the release of natural or bioengineered strains of smallpox as a biologic weapon. In the U.S., we stopped smallpox vaccination more than 25 years ago (1972). Henderson estimates that only 10 to 15% of the population has residual smallpox immunity.[9] The U.S. Centers for Disease Control and the Russian State Research Center of Virology and Biotechnology have viable variola virus; other unreported or bioengineered strains may exist. The potential release of variola with its high mortality (>25%) and high transmissibility (up to 50%) could be catastrophic because of inadequate vaccine stores and production capacity and lack of effective treatment. Some now recommend that those who would be involved in response to a bioterrorist act should be vaccinated.[7]

The potential for mass casualty and significant spread from a point source is exemplified by the history of smallpox spread in the New World during the age of exploration. Twenty-six years before the first contact of the Stolo society and European explorers, more than two-thirds of the tribe was wiped out by smallpox. The disease was introduced into the indigenous population of Mexico in 1779, more than 2,000 miles from the Stolo villages in British Columbia. Historians suspect that the epidemic spread up the river trading routes through the plains, the mountains and finally into the woodland and coastal tribes. The disease resulted in the loss of more than 60% of the population—corpses were piled in huts and burned. Because most deaths occurred during the peak of hunting and gathering season, many more were affected by starvation during the winter months. Not only was the population devastated, but the death of tribal elders also resulted in a tremendous loss of the oral traditions and culture of these people.[4]

BOVINE TUBERCULOSIS

Tuberculosis is one of the oldest recognized diseases of man and animals. Evidence of tuberculosis was found in bones in ancient Egypt, India, Persia, and pre-Columbian Mesoamerica. The clinical features of tuberculosis (phthisis, scrofula, and gastrointestinal disease) were described by Hippocrates and other early Greeks.[10,11] Post-mortem examinations done by Sylvius in the seventeenth century demonstrated the classic tubercle from which the term tuberculosis is derived. Ancient Judaic teaching from the Talmud warned that any animal carcass showing adhesions between the lungs and the pleura (tuberculosis?) was unsatisfactory for human consumption; suggesting that animal-to-human transmission was recognized at that time. There is paleologic evidence that llamas, dogs, and bison may have served as reservoirs of tuberculosis in the New World.[11]

In 1882, Robert Koch identified the bacillus (*Mycobacterium tuberculosis*) that caused human tuberculosis. He showed that it could be transmitted to animals. Koch believed that the organisms in humans and cattle were the same, but Theobald Smith, a veterinarian from Harvard, demonstrated that they were different organisms.[11] Most species of mycobacteria can be transmitted across animal species, but not all infections cross species with the same efficiency. The three main types identified were human, bovine, and avian. Horses, mules, sheep, and goats are resistant to all three types, but swine are susceptible to all three (most get avian). Cattle are susceptible to all three but only slightly to human and avian strains. Dogs can get human, cats can get bovine, but chickens get only avian. Immunocompetent humans are resistant to *M. bovis*, but severe or recurrent exposure in milkers, cattle workers, and milk drinkers can cause disease.[12]

Lister, Bang, and McFadyeann presented data at a conference in 1901 showing the relationship between bovine tuberculosis, infected milk, and human disease. At that time, hundreds of children per year died of tuberculous meningitis or miliary tuberculosis from contaminated milk. Others had scrofula or arthritis. Unpasteurized milk of tuberculous cows was the source of more than 90% of infections in children.[11,12] These findings led to campaigns in several countries to eradicate bovine tuberculosis.

Unlike smallpox, bovine tuberculosis (1) often does not produce acute illness; (2) has a carrier stage; (3) has multiple non-human reservoirs; (4) has no vaccine; (5) has no centralized global surveillance or eradication programs; and (6) control measures have significant economic consequences. Prevention and control are costly and labor intensive: testing of herds with branding and slaughter of reactive animals, disinfecting stock areas after eradication of reactors (especially manure), educating stock handlers to test all new stock before integration in the herd, pasteurizing milk, keeping cattle from potentially contaminated streams and grazing areas, and limiting exposure during shipping, in stockyards, and at shows (cattle tested three months post-exposure).[12] These factors have presented major historical and present day obstacles to eradication of bovine tuberculosis.

The United States had one of the earliest, most extensive eradication programs—it is also one of the most successful. In May 1917, the U.S. Bureau of Animal Industry initiated a federal-state program for the eradication of bovine tuberculosis based on the test-and-slaughter plan with payment of indemnity for animals destroyed. State and federal inspectors followed rigorous guidelines and cattle owners failing to comply were punished:[12]

> It was unlawful a) "to obstruct, attack or interfere with...anyone testing cattle; b) attempt to defeat, obstruct or interfere with application of tuberculin test" In order to carry out testing, inspectors had the right "at any time [to] enter any premises, except dwelling houses" Failure to comply with the regulations allowed the Sheriff to carry out required testing, branding and/or slaughter with the authority to place a lien on the cattle. Violators would be "guilty of a misdemeanor" and "punished by a fine of not more than $500, or by imprisonment in the county jail not exceeding six months, or by both..." (Ord. 225 §10-13, 1935, Marin County California).

Cattle from 3,017 counties in the U.S., Puerto Rico, and Virgin Islands had been tested. This required 232 million tuberculin tests and slaughter of 3.8 million animals (with repayment to the owners). The greatest prevalence was in northern dairy states with rates of 40–80% in the some large dairies supplying milk products. The relentless campaign of eradication reduced the number of tuberculous cattle in the United States from one animal in every 20 to less than one in 200 by the 1940s. The Meat Inspection Division of the Bureau of Animal Industry found 0.53% of meat contaminated in 1917; by 1941 the prevalence was reduced to 0.02%.[12] This marked decrease in prevalence of bovine tuberculosis increased the rates of susceptibility, requiring surveillance by testing and meat inspection to completely eradicate the disease and prevent resurgence. This surveillance system is still in place today.

A cost-benefit analysis of *M. bovis* eradication in the U.S. showed an actual cost of $538 million between 1917 and 1992; $255 million federal and $283 million state funds were used. The current programs cost approximately $3.5 to $4 million per year. By reducing the number of cattle lost from 100,000 head to less than 30 per year, the program saves $150 million per year in replacement costs alone.[14,15] In addition to the direct savings, farmers eliminated indirect cost of losses in milk and meat production, stock replacement, and decontamination procedures.

Human mortality from tuberculosis also dropped from 150 per 100,000 population in 1918 to $\ll 50$ per 100,000 in 1942.[12] The introduction of streptomycin, isoniazid, and para-amino salicylic acid (PAS) in the 1940s and 1950s lowered rates of tuberculosis to less than five per 100,000 population by 1980.[10] At that time, approximately 0.3% of human tuberculosis was due to *M. bovis*. Because of HIV-associated immunosuppression and other factors, tuberculosis rates in the U.S. increased more than 20% from 1986 to 1990. *M. bovis* caused 3% of cases.[10,13]

Many countries have not had the same level of success. Infected milk may have played a role in the spread of tuberculosis in South Africa in the early 1900s. In 1905, one veterinarian reported that up to 60% of dairy herds in the Cape provinces were reactors; 12% were producing infected milk. Before that time it was assumed that *M. bovis* did not exist in South Africa. The Cape Act No. 16 (1906) allowed for testing and slaughter of cattle with compensation of only one-fourth the value of the animal to the farmer. At that time 16,796 animals were tested: 2.5% had a positive reaction and had to be slaughtered. In an attempt to limit the spread of infection, the transport of cows from provinces with higher rates of infection was prohibited. Economic factors, however, were allowed to override these policies. Rather than incurring loss, some owners "dumped" known reactors in Natal and Transvaal provinces where dairymen were often naïve about bovine tuberculosis. If tuberculosis was reported in slaughtered cattle, the government conducted testing and slaughter of infected cattle from the home herd. To avoid these potential losses, many farmers stopped the shipment of cattle to the public abattoir.[16]

Testing and slaughter of dairy herds that provided milk to urban centers resulted in severe economic losses, causing the farmers to pressure the provincial council to stop slaughter of "valuable cattle" and "prevent the destruction of the dairy industry."[16] Since it was too costly to give full compensation, the government decided to implement voluntary testing. As a result, dairymen failed to report outbreaks in their herds and actually desensitized animals to tuberculin to get false negative reactions. Despite ineffective attempts to control the problem, no serious program was implemented until after World War II. The only option available to the health authorities was to advocate pasteurization, but as late as 1954 only half the milk of Johannesburg was heat-treated.[16]

Bovine tuberculosis is uncommon in wild animals not exposed to infected domestic stock or humans, but is well known in animal populations that share watering holes or pastureland with infected livestock.[11] Recent Promed listings report cases of *M. bovis* in wildlife throughout the world. Cases include in deer, bison, otters, squirrels, coyotes, and raccoons in the U.S. and Canada; camels in Egypt; Cape buffalo, giraffes, antelope, elephants, and predatory felines in several African nations; badgers in the U.K.; and opossums in New Zealand. Infection in predatory felines has led to a significant loss of lions in Kruger National Park, South Africa. Half of some herds of wood bison in Canadian parks are tuberculin reactors. Wolf predators are not infected (probably due to natural resistance in canines) but scavengers such as coyotes and raccoons are. A study in Michigan on *M. bovis* isolates from a cow, four coyotes, two raccoons, three captive white-tailed deer, and 84 free-ranging white-tailed deer from affected areas revealed very similar DNA fingerprints, suggesting cross-species transmission.[17]

The issue of transmission between domestic livestock and wild animal populations adds yet another obstacle to eradication efforts as illustrated by the badger problem in the United Kingdom. Following the eradication program, only 0.06% of cattle in the U.K. were PPD reactors. Most cases were from southwestern England where high rates of *M. bovis* were found in wild badgers. Epidemiology studies done in 1965 revealed that areas with higher badger densities had significantly higher rates of *M. bovis* in local cattle herds.[18] In August 1998, the problem continues; the incidence in cattle increased in the West region from 316 cases five years ago to more than 500 in 1998. Badgers are an endangered species and therefore protected from slaughter, but the government is also responsible for public safety and protection of livestock.[17]

A total of six million pounds sterling/year has been set aside for implementing a five-point strategy to deal with this issue: (1) minimize risk to humans, (2) develop vaccine for cattle, (3) conduct studies to better understand interspecies transmission, (4) prevent cattle-to-cattle spread (test and slaughter with 100% compensation), and (5) badger culling. The program hopes to satisfy wildlife advocates, dairy farmers, and public health officials. In addition, pasteurization and rigorous meat inspection are required to prevent transmission to humans.[17]

PAHO announced at the 1997 Tenth InterAmerican Ministerial Meeting on Animal Health that the prevalence of *M. bovis* in cattle was less than 1% in Central America and was in the process of eradication in Uruguay, Chile, Columbia, Argentina, Paraguay, and Venezuela.[14] Those countries with inadequate control programs have 24% of the region's cattle and 60% of the total population. In Africa and other developing countries, data are scarce and there are tremendous resource constraints on control activities. As much as 90% of the human population in Africa live in countries where cattle and dairy cows undergo no or only partial control. In many areas, pasteurization of locally produced milk is not feasible[13]—up to 90% of milk is consumed fresh or soured.[19] Cattle testing and meat inspections are often not conducted due to lack of trained personnel, lack of resources, or fear of economic loss. The situation is similar in Asia where cattle and buffalo are the principal domestic animals susceptible to *M. bovis*. Ninety-four per cent of the Asian population lives in countries where no or limited control programs exist. Unlike the U.S. where less than 5% of the population has regular, direct exposure to cattle, more than 50% of people in Asia and Africa have close contact with potentially infected animals.[13] Thus the threat of *M. bovis* is from natural, insidious spread in areas of inadequate surveillance rather than as an agent of bioterrorism.

CONCLUSION

Both human and animal diseases have significant direct and indirect costs. Eradication provides the ideal means of disease control, but is not always feasible. The following questions should be asked when contemplating an eradication effort: Is eradication possible? Are there potential biologic, environmental, social, cultural, and economic obstacles? What are the direct and indirect costs—both short and long term? What are the costs of doing nothing?

The smallpox and bovine tuberculosis programs illustrate many of the factors that influence the implementation and outcome of eradication efforts. Although the partial eradication of bovine tuberculosis in a single country took 40 years and $100 million dollars more than the complete eradication of smallpox from the globe, both programs paid for themselves within three years of completion. Smallpox has become a potential bioterrorist threat to the human population because of the complete success of the program. *M. bovis* remains a real threat to agricultural and wildlife because of the uneven success of its eradication. For both, commitment of human and financial resources to active surveillance and public education continue to be essential for prevention and control.

The threat of zoonotic orthopoxvirus infection and transmission of bovine tuberculosis between wild and domestic animals is increased by current rural agricultural practices. Recent cultural and social changes have moved the right of the individual (animal or human) ahead of public safety. Issues of public perception and political pressure influence budget decisions. Decreasing funds significantly limit response capabilities of health and agriculture agencies, at the same time the uneven distribution of wealth around the globe, the ease of travel, and the increasing import/export food markets have increased the potential of disease transmission.

REFERENCES

1. GLICKMAN, L. & N. GLICKMAN. 1998. The epidemiology of human-animal interactions. Part I. Zoonotic Diseases. pitt.edu/~super1/lecture/lec0301.
2. STRANO, A.J. 1976. Smallpox. *In* Pathology of Tropical and Extraordinary Diseases. C. H. Binford & D. H. Connor, Eds.: 65–67. AFIP. Washington D.C.
3. FENNER, F. 1994. Poxviral Zoonoses. *In* Handbook of Zoonose. 2nd edit. Section B: Viral. G.W. Beran & J. H. Steele, Eds.: 485–503. CRC Press, Inc. Boca Raton, FL.
4. CARLSON, K.T. 1998. First Contact: Smallpox. web20.mindlink.net/stolo/firstcon.
5. HENDERSON, D.A. 1976. The eradication of smallpox. Sci. Am. **235:** 25–33.
6. BREMAN, J.G. & I. ARITA. 1980. The confirmation and maintenance of smallpox eradication. N. Engl. J. Med. **303:** 1263–1275.
7. BREMAN, J.G. & D.A. HENDERSON. 1998. Poxvirus dilemmas—Monkeypox, smallpox, and biologic terrorism. (Sounding Board) N. Engl. J. Med. **339:** 556–559.
8. BREMAN, J.G. 1991. Viral infections with cutaneous lesions. Poxviruses: Variola, vaccinia, monkeypox, tanapox. *In* Hunter's Tropical Medicine. 7th edit. G.T. Strickland, Ed.: 167–170. W.B. Saunders. Philadelphia, PA.
9. HENDERSON, D.A. 1998. Bioterrorism as a public health threat. J. Am. Med. Assoc. **4:** 488–492.
10. BLUMBERG, H.M. & A.M. NELSON. 1998. Tuberculosis. *In* Pathology of Emerging Infections. 2. A.M. Nelson & C.R. Horsburgh, Eds.: 167–192. ASM Press. Washington, D.C.
11. THOEN, C.O. & D.E. WILLIAMS. 1994. Tuberculosis, tuberculoidosis and other mycobacterial infections. *In* Handbook of Zoonoses. 2nd edit. G.W. Beran, Ed.: 41–59. CRC Press. Boca Raton, FL.
12. WIGHT, A.E., E. LASH, H.M. O'REAR *et al.* 1942. Tuberculosis and its eradication. *In* Keeping Livestock Healthy. Yearbook of Agriculture. U.S. Department of Agriculture. 77th Congress, House Document 527. General Printing Office. Washington, D.C.

13. COSIVI, O., J.M. GRANGE, C.J. DABORN *et al.* 1998. Zoonotic tuberculosis due to *Mycobacterium bovis* in developing countries. Emerging Infectious Diseases (Synopses). January-March: 59–70.
14. GUIDE FOR BOVINE TUBERCULOSIS PROJECTS. 1997. Guidelines for the preparation of plans for programs of bovine tuberculosis eradication and principles and technical criteria for the conduct and evaluation of bovine tuberculosis eradication programs. PAHO Technical Note N. 15/Rev.2, March, 1997.
15. FRYE, G. 1994. Bovine tuberculosis eradication. *In Mycobacterium bovis* Infection in Humans and Animals. C.O. Thoen & J.H. Steele, Eds. Iowa State University Press. Ames, IA.
16. PACKARD, R.M. 1989. White Plague, Black Labor. p. 42–48. University of Natal Press. Peitermaritzburg.
17. PRO/AH>Mycobacterium bovis, wildlife. 1998 promed@usa.healthnet.org
18. WILESMITH, J.W. & M.S. RICHARDS. 1992. Bovine tuberculosis in cattle and badgers. Third International Symposium on Veterinary Epidemiology and Economics. p. 590–597. Veterinary Medicine Publishing Co. Kansas.
19. WALSHE, M.J., J. GRINDLE, A. NELL *et al.* 1991. Dairy development in sub-Saharan Africa. African Technical Department Series. World Bank Technical Paper No. 135. World Bank. Washington, D.C.

Economic Considerations of Agricultural Diseases

CORRIE BROWN

Department of Pathology, College of Veterinary Medicine, University of Georgia, Athens, Georgia 30602-7388, USA

The U.S. livestock industry is among the most economically viable in the world, largely as a result of consistently excellent health status. This is due in no small measure to concerted efforts to exclude diseases present in many other areas of the world. The United States is free of most Office International des Epizooties "List A" diseases, all of which are so classified based on their potential for rapid spread and resulting socioeconomic consequences. The paucity of disease in the United States translates into increased productivity of our national herds, which in turn means lower prices for the consumer and greater profits for the producer. The American consumer pays the lowest percentage for food of any country in the world, approximately 12 cents/dollar[1] earned whereas in many other countries, the cost is as high as 50 or 60 cents/dollar. Another index of economic viability is the value of American livestock in the international marketplace. Many sectors of agriculture are driven by their ability to export and because of the general lack of disease, U.S. animals and animal-product export markets are very attractive and generate considerable economic transactions.

All of this economic vitality is dependent on freedom from disease. A bioterrorist event could change the disease status of our national herd in a precipitous way with devastating results. The American public is generally unaware of this potential for economic ruin. To raise awareness in our public-policy makers on this issue will require a concerted and targeted effort.

Whenever an unexpected disease enters the United States, both consumer and export markets are negatively affected. That is, a spreading disease increases prices at the supermarket, resulting in a pinch for the consumer and a simultaneous drop in export-market transactions. A brief examination of some of the literature concerning economics of foreign animal diseases provides insight into how a bioterrorism event could hit us square in the pocketbook and cause losses exponentially greater than what the general public might expect.[2]

The last major foreign animal disease outbreak in the United States was a highly pathogenic avian influenza in 1983–1984. This outbreak was confined to a relatively small area encompassing parts of Pennsylvania and neighboring states, but nevertheless was the most costly and extensive eradication effort in our history. In six months, all infected chickens were depopulated and premises decontaminated, with a price tag of US$63 million paid by the federal government. Despite the seemingly exorbitant cost, the decision to carry out the eradication was made easily, as economic analyses demonstrated that in the absence of eradication, the final cost of living with highly pathogenic avian influenza would have been US$5.6 billion, all passed on to American consumers in the form of increased meat and egg prices.[3] As it was, during the six months of the outbreak, poultry prices increased by US$349 million.[3]

Similarly, a study done on the cost of maintaining a hog population with African swine fever, a disease that is endemic in Africa and has been present periodically in the Caribbean, revealed surprising hidden costs. The cost, over a ten-year period, would be US$5.4 billion, with the bulk of this being consumer losses.[4]

The disease that causes the greatest concern among producers and regulators is foot-and-mouth disease. This is an extremely contagious viral disease affecting a wide range of animals, including pigs, cattle, sheep, goats, and many species of wildlife. Spread by aerosol, the virus is capable of almost uncontrollable spread. The virus grows in epithelium of the oral cavity and the feet where, in both cases, it causes painful blisters, which make the animal unwilling to eat or move around to forage. As a result, there is a tremendous, albeit temporary, drop in production. Such a drop, however brief, would be enough to wipe out profits in our current systems of intensive agriculture. A study done almost twenty years ago still provides good data for analysis of economic impact of this disease.[5] In this study, it was hypothesized that foot-and-mouth disease entered the United States and became endemic. Efforts to control the disease were unsuccessful. Export losses were the largest negative consequence, with numbers adjusted for inflation registering US$27 billion in lost trade. Recent outbreaks of foot-and-mouth disease around the world have underscored the potential of this disease to hamper economic development. Foot-and-mouth disease was transported to Italy in 1993, at which time Italy had been free of the disease for four years. To control the outbreak, 8,000 animals were slaughtered and eradication costs were estimated at US$8.3 million for indemnity funds and US$3.2 million for cleaning, disinfection, and carcass disposal. However, indirect costs, as measured by disruption of international trade, were estimated at US$120 million, or approximately ten times the cost of depopulation and decontamination.[6] Another sobering example exists with the outbreak of foot-and-mouth disease in Taiwan in 1996. To date, more than 3.8 million hogs have been slaughtered and early estimates of losses to the swine-related industries in Taiwan are approximately US$7 billion.[7]

A disease with zoonotic potential will have an even greater impact. Economic figures released by Great Britain regarding impact of bovine spongiform encephalopathy (BSE) are staggering. The value of British beef and beef products, estimated at US$880 million, fell considerably when, in 1988, BSE was declared as a newly emerging disease problem of cattle. However, the value fell fully to zero in March of 1996 when it was announced that there was a probable link between consumption of BSE-affected meat and new variant Creutzfeldt-Jakob in humans. The direct costs of dealing with the outbreak are staggering. A regulatory decision mandated slaughter of all cattle over the age of 30 months. As a result, approximately 1.35 million were destroyed and all of the carcasses disposed of by incineration. This depopulation cost surpassed US$4.2 billion and continued to climb.[8] Even a threat of placing prions in our food supply would send a staggering negative ripple through the beef and dairy industries.

The possibilities of obtaining and transporting infectious materials such as those listed above are relatively straightforward. Visiting an area where one of these dreaded diseases is endemic, coupled with a rudimentary knowledge of microbiology, could be enough to allow for production of a bioterrorist weapon that could then be released in naive populations in the United States. Our only defense against such an episode is to increase awareness to a point where such an incursion is detected as early as possible and deleterious spread effectively intercepted.

REFERENCES

1. U.S. DEPARTMENT OF COMMERCE. 1994. Statistical abstract of the United States. U.S. Department of Commerce. Washington, D.C.
2. BROWN, C.C. & B.D. SLENNING. 1996. Impact and risk of foreign animal diseases. J. Am. Vet. Med. Assoc. **208:** 1038–1040.
3. LASLEY, F.A., S.D. SHORT & W.L. HENSON. 1985. Economic assessment of the 1983–1984 avian influenza eradication program. ERS Staff Report No. AGES841212. National Economics Division, Economic Research Service. U.S. Department of Agriculture. Washington, D.C.
4. RENDLEMAN, C.M. & F.J. SPINELLI. 1994. An economic assessment of the costs and benefits of African swine fever prevention. Animal Health Insight. Spring/Summer.
5. MCCAULEY, E.H. *et al.* 1979. A study of the potential economic impact of foot-and-mouth disease in the United States. U.S. Government Printing Office. Washington, D.C.
6. TANAKA, R. 1993. Foot-and-mouth disease in Italy. Foreign Animal Disease Report. USDA-APHIS-VS-EP **21**(2/3)**:** 8–9.
7. WILSON, T.M. & C. TUSZYNSKI. 1998. Foot and mouth disease in Taiwan—1997 Overview. Foreign Animal Disease Report. USDA-APHIS-VS-EP. Summer 1998.
8. ANONYMOUS. 1998. £2.5 billion and rising. Vet. Rec. **143:** 57.

Regionalization's Potential in Mitigating Trade Losses Related to Livestock Disease Entry

ANN HILLBERG SEITZINGER,[a,d] KENNETH W. FORSYTHE, JR.,[b] AND MARY LISA MADELL[c]

[a]*Centers for Epidemiology and Animal Health, Veterinary Services, Animal and Plant Health Inspection Service, U.S. Department of Agriculture, Fort Collins, Colorado 80521, USA*

[b]*Centers for Epidemiology and Animal Health, Fort Collins, Colorado 80521, USA*

[c]*Trade Support Team, International Services, Animal and Plant Health Inspection Service, U.S. Department of Agriculture, Washington, D.C. 20250, USA*

INTRODUCTION

Entry of plant and livestock pests and disease agents may lead to significant production and trade losses and to high control costs. Regionalization offers the potential for limiting the trade losses as countries recognize areas of low pest and disease transmission risk that may not correspond to political boundaries. Trade may be allowed from these areas, or "regions," but may not be from other areas of the country posing higher risks of transmission. The concept of regionalization is fairly simple and intuitive, and estimated potential gains of implementation have been large. While implementation of regionalization requires considerable and timely forethought, it must be considered a desirable tool when it is neither possible nor desirable to implement policies imposing zero risk of pest or disease entry.

APPROACHES TO REGIONALIZATION

There are two approaches to regionalization. Neither is new. One approach effectively defines regions by the absence of pests or diseases, while the second approach defines regions according to the presence of pests or diseases (FIG. 1). If the absence or presence of pests or diseases could be determined with certainty, the two approaches toward defining regions would lead to the same endpoint at a given point in time. Without this certainty, the approaches will probably not converge, but they may be considered complimentary in terms of achieving the maximum benefits from regionalization. Both approaches have been accepted in the international arena with the burden of proof for a regionalization scheme's efficacy placed on the exporting country.

---

[d]Address correspondence to: Ann Hillberg Seitzinger, Centers for Epidemiology and Animal Health, Veterinary Services, Animal and Plant Health Inspection Service, U.S. Department of Agriculture, 555 South Howes, Fort Collins, Colorado 80521; Telephone: 970-490-7843; Fax: 970-490-7899.

e-mail: Ann.H.Seitzinger@usda.gov

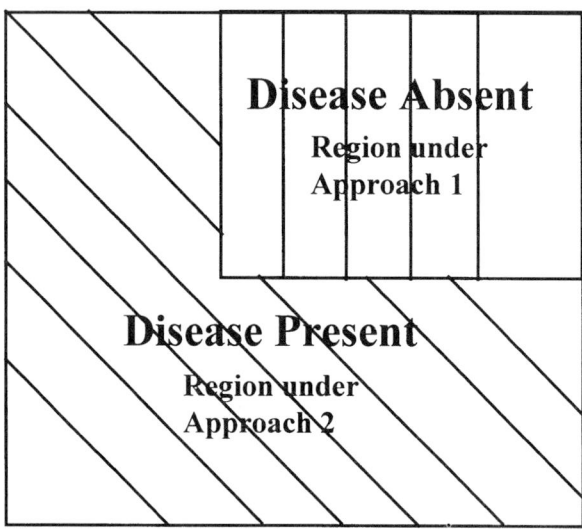

FIGURE 1. Two approaches to regionalization.

Defining Regions by the Absence of Disease

Defining regions by the absence of disease provides a means for lower risk regions to enjoy the benefits of facing lower trade barriers to their animals, plants, and products. Two examples of this approach to regionalization involve live cattle trade between the United States and Canada. Canada recognizes differing risk levels for Bluetongue in the United States based on varying incidence levels of disease between the northern, middle, and southern states. This is linked to the apparent absence of competent vectors for the disease in northern states. Proposed changes further reducing the Bluetongue testing requirements on U.S. cattle being exported to Canada in 1994 give an indication of the benefits even limited implementation of this approach to regionalization offers. The elimination of a requirement for two negative Bluetongue tests on breeding cattle from low incidence states during a vector-free period and the reductions to only one negative test for breeding cattle from high incidence states during the vector-free period and for feeder cattle from medium incidence states during the vector period was estimated to reduce testing costs by between $298,000 and $334,000.

A second example of the application of this type of regionalization is the Northwest Pilot Project, implemented in 1997. Testing requirements for anaplasmosis, brucellosis, and tuberculosis have been eliminated for U.S. cattle exported to Canada from the low disease risk states of Montana and Washington. Young and Marsh[1] estimate that the removal of these testing requirements leads to a $27 per head reduction in marketing costs. This reduction in marketing costs is estimated in turn to lead to a $0.14/hundredweight increase in the U.S. feeder price, unadjusted for supply response.

Defining Regions by the Presence of Disease

Defining regions based on the presence of disease, especially in the case of infrequent outbreaks, allows unaffected regions to continue trading their goods. Two examples of this approach to regionalization involve Canada's responses to an outbreak of highly pathogenic avian influenza in 1983 in the northeastern United States and to an outbreak of vesicular stomatitis virus in 1995 in the U.S. mountain states. In 1983, Canada continued to import poultry products from unaffected states. Recent estimates of the benefits of all trading partners continuing to import poultry products under this type of regionalization approach would limit losses in U.S. broiler farm income to an estimated $13 million as opposed to the $351 million estimated decline in broiler farm income projected should all U.S. poultry products exports be embargoed.[2]

Canada's approach to the 1995 U.S. vesicular stomatitis outbreak was to limit restrictions to horses, ruminants, and swine originating in states that had had cases in the previous 30 days. Once quarantines in an affected state were lifted, Canada required that livestock from that state test negative for serum neutralizing antibodies to vesicular stomatitis virus, be from unaffected premises, and be protected from insect vectors for 30 days before export. The only regulation affecting all trade from the U.S. into Canada involved subjecting all horses entering Canada from the U.S. to inspection by Canadian federal veterinarians upon entry. As a result of these targeted measures total U.S. trade in bovine, equine, and porcine with Canada valued at approximately $43 million was reduced by only negligible amounts.

IMPLEMENTING REGIONALIZATION

Although regionalization has been put into practice in the past, recent commitments under the General Agreement for Tariffs and Trade (GATT) and the North American Free Trade Agreement (NAFTA) sanitary and phytosanitary provisions have enhanced countries' potential for implementing regionalization. In response to these commitments, the U.S. Department of Agriculture's (USDA) Animal and Plant Health Inspection Service (APHIS) has adopted a regionalization policy to be applied to regions of countries wishing to export to the United States.[3] The implementation of this policy involves evaluating the risks presented by proposed animal and animal-product importations based on a range of characteristics of the region from which they are exported, rather than on disease-free or affected statuses determined on a country-by-country basis.

With minor exceptions, the portions of the Code of Federal Regulations (CFR) dealing with animal and animal-product imports were based previously on assigning a single disease status to the entire region defined by a country's national political boundaries. The new APHIS policy recognizes that, for the purpose of evaluating disease risk, a region may be defined as any geographic land area identifiable by geological, political, or surveyed boundaries. In other words, within a single country there may be many regions that have different risk characteristics that would necessitate the imposition of different U.S. import requirements.

Countries have typically been classified in the CFR as disease-free or affected. However, there has been some recognition previously in the CFR of the range of

TABLE 1. Factors in exporting regions' veterinary services

Disease surveillance
Diagnostic laboratory capabilities
Disease status
Active disease control
Vaccination status
Disease status of adjacent regions
Separation from higher-risk regions
Import restrictions
Livestock demographics
Emergency response capability

risks that countries may present through an additional classification known as modified-free in cases where a disease agent may exist in close proximity to an otherwise free country or where the import practices of the country in question suggest a slightly higher level of risk. In general, therefore, the CFR has operated under a three-category risk classification system, where countries are classified as free, modified-free, or affected.

The new APHIS policy expands the recognition of the range of risks that may be presented by regions by outlining five benchmark risk levels. These are negligible, slight, low, moderate, and high risk. The negligible, slight, and high risk benchmarks correspond roughly to the free, modified-free, and affected classifications previously applied in the CFR. The low and moderate risk benchmarks represent levels of risk not generally recognized in the CFR. The low risk benchmark may be thought of as a free-with-vaccination benchmark, and the moderate risk benchmark may be thought of as regions where the disease agent is known to exist but at a low prevalence. The free-with-vaccination benchmark has recently been applied to help implement the 20,000 metric ton tariff-rate quota negotiated for U.S. beef imports from Argentina. These imports could not have been accomplished under the three-category system previously employed.

In the interests of transparency regarding the approach to evaluating the risk presented by different regions, the APHIS policy statement identifies the specific factors for which the exporters must provide data and information about their country (TABLE 1). These data and information are critical for assessing the capabilities of the exporting country to identify and control pests and disease. They are essential components for estimating the likelihood of the exporting country introducing a particular pest or disease into the United States.

The APHIS policy statement also includes some guidance regarding how APHIS evaluates these factors in relation to the five benchmarks. For example, under negligible risk, vaccination would not be permitted nor could vaccinated animals be present; while under low risk, vaccination would be limited to the herds at greatest risk of exposure. The policy statement is intended to provide some insight into how APHIS evaluates risk but does not imply a rigid classification scheme based on the five benchmarks. All requests from potential exporting regions will be evaluated on a case-by-case basis and each case will be evaluated based on the merits of the individual situation. Based on the evaluation, APHIS may develop a proposal regarding the appropriate import requirements that would be needed to present a negligible level of risk to the United States. The proposal would be published in the Federal Register to provide an opportunity for the public to review and comment on the evaluation and the proposal and to provide any additional relevant input to APHIS.

The APHIS regionalization policy for livestock and livestock products appears to offer the potential for application of both approaches to regionalization. However, the extensive data and information requirements necessary to verify the efficacy of a scheme indicate the considerable and timely effort an exporting country must exert to maintain regionalization as a viable option, particularly in the event of an outbreak.

CONCLUSION

In attempting to address the risk of foreign animal disease entry, whether accidental or through terrorist activity, the temptation is to drive the risk toward zero by eliminating all movement of plants, animals, and their products. However, economic analyses show that trade offers gains when it occurs and exacts costs when it does not occur due to intervention. Regionalization of areas according to the risk of pest or disease transmission offers the potential for greater gains from trade to be realized by directly recognizing the risk of pest or disease transmission.

REFERENCES

1. YOUNG, L.M. & J.M. MARSH. 1997. Live Cattle Trade between the United States and Canada: Effects of Canadian Slaughter Capacity and Health Regulations. Montana State University Trade Research Center Research Discussion Paper No. 7. Bozeman, MT.
2. DISNEY, W.T. & M.A. PETERS. 1994. Economic implications from U.S. regionalization of broiler exports under quarantine conditions. Presented at the Southern Agricultural Economics Association Meetings. Nashville, TN, February 5–9, 1994.
3. October 28, 1997. Importation of Animals and Animal Products; APHIS Policy Regarding Importation of Animals and Animal Products; Final Rule and Notice. Fed. Reg. **62**(208): 55999–56026.

Foreign Animal Disease Agents as Weapons in Biological Warfare

DAVID R. FRANZ[a]

Chemical and Biological Defense Division, Southern Research Institute, 365 West Patrick Street, Suite 223, Frederick, Maryland 21701-4856, USA

Biological warfare may be defined as the intentional use of microorganisms or toxins derived from living organisms to produce death or disease in humans, animals, or plants. For nearly 50 years, world powers have developed biological weapons and countermeasures to biological weapons. The focus of biological defense programs has been protection of humans in the context of military forces on the battlefield. Offensive biowarfare programs traditionally select anti-human agents based on the physical and biological characteristics that maximize their utility as weapons of war. Important characteristics include pathogenicity or toxicity, ease of production, and stability—stability during development, production, storage, and use on the battlefield. Because, unlike most chemical warfare agents, bacteria, viruses, and toxins are neither volatile nor dermally active, it was necessary to select agents that could be presented as respirable aerosols (1–25 µm particles) to be inhaled by the human target. Years of primarily independent research by five to six proliferent nations resulted in essentially the same list of 10–20 microbial agents. These agents, if properly prepared and delivered under ideal meteorological conditions, could infect unsuspecting humans without their knowledge and ultimately lead to disease or death of thousands within days or a few weeks (TABLE 1).[1] It is important to note that with the exception of smallpox, which was eradicated as a natural disease in the 1970s[2] the threat agents selected for use against humans are not highly contagious. That is, transmission from an individual already ill to a healthy individual through normal contact is unlikely. Therefore, to maximize impact agent selection was limited to those with the physical and biological characteristics allowing generation of an aerosol cloud. The use of highly contagious agents, such as smallpox and highly pathogenic influenza strains, as weapons could preclude the need to develop complex research and development programs, scale up and production schemes, and expensive delivery systems. Simply introducing a few infected individuals into a naive population could start a devastating epidemic, or possibly even a pandemic.

The U.S. response to the cold-war biological threat was multifaceted: detection and physical protection, medical defense, intelligence, active defense and threat reduction, and treaties and nonproliferation efforts. Specific countermeasures to protect forces on the battlefield include active immunization, passive immunoprophylaxis and chemoprophylaxis, battlefield detection, physical protection, identification and diagnosis, decontamination, and passive immunotherapy or chemotherapy. These

---

[a]Address correspondence to: David R. Franz, D.V.M., Ph.D., Chemical and Biological Defense Divison, Southern Research Institute, 365 West Patrick Street, Suite 223, Frederick, Maryland 21701-4856; Telephone: 301-668-6141; Fax: 301-668-6146.
e-mail: franz@sri.org

TABLE 1. Hypothetical human health impact of dissemination by airplane of 50 kg of agent along a 2-km line upwind of a population center of 500,000 humans

Agent	Downwind Reach (km)	Dead	Incapacitated
Rift Valley fever	1	400	35,000
Tick-borne encephalitis	1	9,500	35,000
Typhus	5	19,000	85,000
Brucellosis	10	500	100,000
Q fever	>20	150	125,000
Tularemia	>20	30,000	125,000
Anthrax	≫20	95,000	125,000

NOTE: The agents listed are typical of those weaponized by proliferators during the mid-twentieth century and are not highly contagious, and therefore require respirable aerosol dissemination. (Adapted from World Health Organization.[1])

measures are appropriate for reducing or eliminating the threat of biological agent use against our forces, and probably serve as an effective deterrent against its use in war.

In the last 10 years of the twentieth century, technological and political factors have changed the face of biological warfare. Biotechnology now makes genetic manipulation of bacteria and viruses possible in even the most scientifically undeveloped nations. Designer agents might be constructed that could expand the old threat lists. At the same time, the demonstrated superiority of the coalition force against Iraq in 1991 has led third-world adversaries and nonstate terrorist groups to consider unconventional weapons as a means of leveling the playing field against the remaining superpower. Finally, it is possible that the dissolution of the Soviet Union with its massive biological warfare infrastructure might actually fuel the programs of these would-be proliferators through technology transfer and "brain drain."

The terrorist, to bring biological warfare to our cities, must play by generally the same technical rules that constrained cold-war proliferators—and typically without the depth of technical infrastructure possessed by the nations. Concern about the threat of biological terrorism raises new challenges and imposes limitations on our treatment capabilities that we did not foresee in the cold-war era. However, the terrorist has a much larger—although not necessarily more lethal—arms chest of agents from which to choose. Even a non-mass casualty attack is enough to trigger extensive press coverage and paralyze a city for many hours. Additionally, because the population at risk is almost infinite and warning extremely unlikely, we can only respond; prophylaxis of the entire population is presently not cost effective. For technical reasons, it is likely that the terrorist will need state sponsorship, the classical cold war agents, and large-scale dissemination techniques in order to kill or infect thousands or hundreds of thousands. Therefore, biological hoaxes or even large-scale, nonlethal foodborne disease agent[3] attacks by terrorists should be more likely than mass casualty, highly lethal attacks—unless very contagious agents are used (FIG. 1). It should be noted, with regard to human disease, that very few highly contagious agents exist.

While biological agents have been used against military animals[4] and weapons have been developed against domestic animals[5] in the past, modern biological pro-

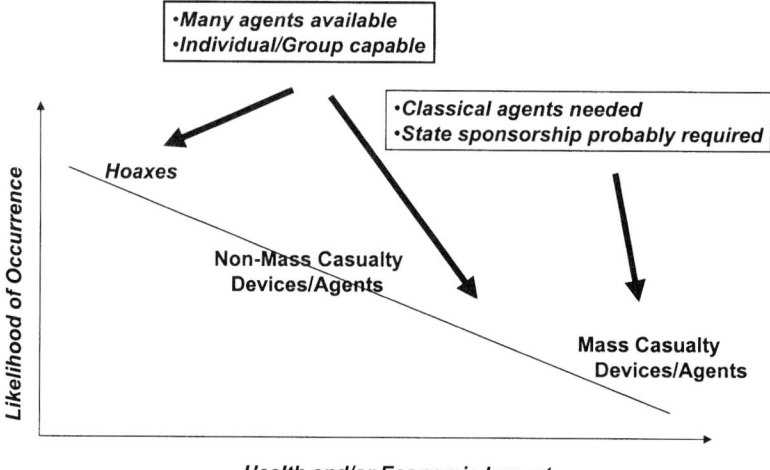

FIGURE 1. Hypothetical spectrum of terrorist attack possibilities against humans comparing likelihood of occurrence—based on ease of execution—with potential impact. Hoaxes or non-mass casualty attacks could have significant political impact. A true mass-casualty attack against a human population would likely require use of the classical agents and necessitate state sponsorship, unless a highly contagious agent such as smallpox were used.

liferators and defensive programs have paid little heed to the vulnerability of our animal populations. The agents historically selected for use against animals, like those designed for human targets, were not highly contagious. They were (to be) delivered in animal feed or by injection.

A major difference between human and animal disease agents in nature is the number of highly contagious agents that can infect animals. Most of these have been eradicated from the U.S. livestock populations and vaccines for them are not used routinely. Thus, livestock populations are vulnerable. Modern high-density husbandry methods, livestock sale and transportation practices, and centralized feed supply and distribution systems only add to the potential of animal-to-animal or fomite transmission. The foot-and-mouth disease outbreak that occurred in the swine population of Taiwan in 1996[6,7] demonstrates the extreme vulnerability of such an industry to contagious disease transmission. Although U.S. practices differ significantly, many of the principles observed and lessons learned in the Taiwanese outbreak apply and should be considered as we evaluate the threat and develop a plan to protect our herds and meat-production industry.

It would seem that attack by non-contagious respirable aerosol, the perceived standard for attack on humans during much of the cold war, would not be the *modus operandi* selected by the animal terrorist. Thus inhalation anthrax, the most feared human threat agent, might not be the first choice for attack against the U.S. livestock industry. The impact of such an attack, even on one of our largest feedlots, would probably not justify the technical requirements of preparation, weaponization, and delivery. However, the small-scale release of one of five to six highly contagious

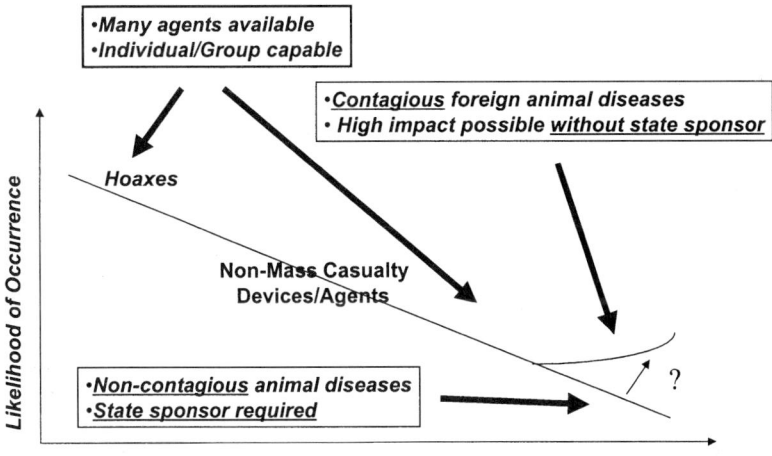

FIGURE 2. Hypothetical spectrum of terrorist attack possibilities against animals comparing likelihood of occurrence—based on ease of execution—with potential impact. As with human disease agents, hoaxes or non-mass casualty attacks on animal populations could have significant economic impact. Use of highly contagious foreign animal disease agents could result in widespread disease and enormous economic impact without a state-sponsored research, development, and weaponization program.

agents would be extremely cost effective for the terrorist and would disrupt our economy and our society (FIG. 2). We can only speculate regarding the relationship between ease of accomplishment and increased likelihood of occurrence.

How might it happen? Following a covert attack with a contagious viral agent on a sale barn or livestock transportation system, local producers might observe an unusual pattern of disease in their herds. Initially, this might appear to be a natural outbreak of an unfamiliar or unknown disease. Local, then state veterinarians, reference laboratories, and veterinary schools might be called in to make the diagnosis. Subsequently, herd diagnoses might be made symptomatically or by field assay, if available. Following an overt attack on single or multiple herds throughout a region, the most urgent need would be to define the extent of the attack and limit the outbreak. Again, reference laboratory and field diagnostics would be the most important tools needed early in the response.

The epidemiological investigation for a covert attack would differ little from that conducted following a naturally occurring outbreak of an animal disease foreign to the United States. The outward presentation of a manmade outbreak would likely differ little from a naturally occurring event. Priorities would be to (1) identify the attack and confirm the agent identification and diagnosis; (2) develop a case definition; (3) identify exposed or potentially exposed herds; (4) control the movement of animals and vehicles out of affected areas; (5) isolate, slaughter, and dispose of (or vaccinate) exposed herds, (6) vaccinate around the outbreak, if possible, and (7) throughout the process, inform and educate the public. Credible informed professionals who can articulate honestly and accurately the nature of the problem and the solutions must be prepared to provide public information. Failure to do so effectively will only worsen the situation and facilitate achievement of terrorist' goals.

The tools exist, in the United States, to deal with the threat of foreign animal disease terrorism. We have veterinary practitioners, state and federal veterinarians, state and federal reference laboratories, and foreign animal disease experts, and federal and military veterinary teams. We have exceptional communication systems and tools for epidemiological investigation. Finally, we have an unprecedented economic interest at stake. As we prepare for the future, we must monitor disease patterns through effective surveillance and herd health programs, teach and practice preventive medicine and biosecurity; educate clinicians, producers, and industry leaders regarding the threat and the agents of concern; develop emergency plans and field diagnostic capabilities; and ever-strengthen our reference laboratory capability.

There is little doubt that terrorist or state actors exist who seek to disrupt our way of life. There is no doubt that an outbreak of a contagious foreign animal disease in our herds would have enormous consequences that would spread throughout our economy. The rules of animal disease terrorism differ from human cold-war biological warfare and not all of the traditional countermeasures are relevant. Because of the contagious nature of a relatively large number of lethal or incapacitating animal diseases that are foreign to U.S. herds, their ready availability in nature outside the United States, and the vulnerability of our livestock industry to attack, we must carefully consider the threat and focus our preparations where they will do the most good. Awareness of the risk, education, effective veterinary preventive medicine programs, and a strong technical research base not only prepare us for the unexpected, but serve as a deterrent by raising the stakes for the would-be foreign animal disease terrorist.

REFERENCES

1. WORLD HEALTH ORGANIZATION. 1970. Health Aspects of Chemical and Biological Weapons: Report of a WHO Group of Consultants. World Health Organization. Geneva, Switzerland.
2. HER MAJESTY'S STATIONERY OFFICE. 1980. Report of the Investigations into the Cause of the 1978 Birmingham Smallpox Occurrence. Her Majesty's Stationery Office. London.
3. TOROK, T.J. et al. 1997. A large community outbreak of salmonellosis caused by intentional contamination of restaurant salad bars. J. Am. Med. Assoc. **278:** 389–395.
4. POUPARD, J.A. & L.A. MILLER. 1992. History of biological warfare: Catapults to capsomeres. Ann. N.Y. Acad. Sci. **666:** 9–18.
5. HARRIS, R. & J. PAXMAN. 1982. A Higher Form of Killing: The secret story of chemical and biological warfare. Hill and Wang. New York.
6. DUNN, C.S. & A.I. DONALDSON. 1997. Foot-and-mouth disease in Taiwan. Vet. Rec. **140**(15): 407.
7. DUNN, C.S. 1997. Natural adaptation to pigs of a Taiwanese isolate of foot–and-mouth disease virus. Vet. Rec. **141**(7): 174–175.

Tools and Methods for Protection of Targets and Infrastructures Associated with Food and Agriculture Industries

DAVID L. HUXSOLL[a]

School of Veterinary Medicine, Louisiana State University, Baton Rouge, Louisiana 70803, USA

Biological warfare can be defined as the use of microorganisms or toxins derived from living organisms to induce death or disease in human beings, animals, or plants. Agents of biological origin may also be used effectively to destroy critical material such as fuels, insulation, and electronics. The definition applies to the lone perpetrator acting independently, to state-supported terrorism, to undeclared wars, and to declared armed conflict. In preparing for an incident, it is logical to address the most severe potential problem and to plan on that basis. The emergency response then could be scaled to meet any situation that might be of lesser magnitude and more confined. In addition to a bioterrorist attack, biological emergencies may result from a number of other events: natural introduction of a new agent; a genetic change in an endemic agent; environmental, social, or economic change; behavioral change of man; epizootics and epidemics secondary to other disasters; and accidental release of a pathogenic agent.

Obviously, the human population is the target of greatest concern. But economic targets, such as livestock, crops, tourism, and transportation, are likely to become terrorist choices in the future. Even small outbreaks of exotic disease in livestock or crops could remove the United States from the large world market which it enjoys for its agricultural products.

During much of the Cold War, concerns in the United States focused on nuclear weapons, both as a threat and as a means of retaliation against attack by a strategic weapon, including chemical and biological. The common response to any mention of a biological threat was, "Nuke 'em!" With emphasis placed on nuclear weapons as a deterrent to attack by any means, issues relating to chemical and biological weapons were often not considered—with the demise of the offensive biological program in 1969 and the signing of the Biological and Toxin Weapons Convention, concern for biological issues all but disappeared. Throughout the 1970s biological warfare issues received little, if any, attention. However, starting in 1979, a series of events heightened the concern over the use of biological agents on the battlefield or in the hands of terrorists.

- The 1979 anthrax outbreak of anthrax in Sverdlovsk, Russia. In 1992, President Boris Yeltsin admitted it was caused by an accidental release from a facility making biological weapons.

---

[a]Address correspondence to: David L. Huxsoll, Office of the Dean, School of Veterinary Medicine, South Stadium Drive, Louisiana State University, Baton Rouge, LA 70803-8402; Telephone: 225-346-3151.

- Yellow rain in Southeast Asia. This issue has not yet been fully resolved.
- Emerging diseases. Since the demise of the U.S. offensive biological weapons program, new diseases with respect to identity or concept have emerged.
- Advances in biotechnology. The field has expanded rapidly since 1970, offering new approaches to the design and development of biological weapons.
- Proliferation of biological weapons programs. We now suspect that at least ten countries have biological warfare programs.
- Breakup of the Soviet Union. The independent countries may maintain or develop biological warfare programs of their own. Additionally, of the thousands of personnel involved in the Soviet Union programs, a number may now seek jobs in other countries.
- The Gulf War. Concerns were high that Iraq would use both chemical and biological weapons. Subsequent United Nations inspections identified a biological warfare program in Iraq.
- The release of the chemical agent sarin in the Tokyo subway system by the Aum Shinrikyo cult in March 1995. Subsequent investigations by the Tokyo police revealed the cult was pursuing the development of biological weapons.
- Recent revelations of the Soviet Union biological warfare program. The size of the program and the nature of the agents in the program have astounded many experts.

In April 1998, the Florida Farm Bureau Federation sponsored a symposium on exotic threats to Florida. The title of the symposium was, "Florida under Attack: Guarding against Biological Threats." The Florida Farm Bureau is commended for the foresight and keen interest in protecting agriculture and food against biological threats that might be naturally or intentionally introduced.

Critical to any response in the event of a biological attack is the rapid identification of the etiologic agent or agents. The causative agent may be microorganisms and toxins that are not normally encountered in the conventional laboratory. This may make the process of their identification slower and more difficult because of the unique reagents, equipment, and specially trained personnel required. Nevertheless, rapid and reliable diagnosis are necessary in view of the critical management decisions that must follow.

Identification of the agent is important in that it will serve to provide information on the number of people or animals that might become ill; an estimate of the duration of illness; the number of people or animals that would likely die or recover; quarantine or isolation requirements; the requirements for treatment and medical support personnel; and means of carcass or corpse disposal.

In protecting potential targets and infrastructures the private sector spends far more money and hires many more people than the public sector. Thus industries of all types that may present a target should be part of the overall effort and plan.

Many lists of potential biological warfare and bioterrorism weapons have been developed. The lists can become quite long. The agent or agents that terrorists might use may be limited only by the competency, ingenuity, and determination of the terrorists. There are several factors that may affect the selection of agents by a terrorist. These include the ease of obtaining a culture, the ease of producing the agent, ease of dissemination, and possibly the publicity associated with the agent. Additionally, it may be quite possible to simply buy the agents from someone working in a biological weapons program in proliferation countries.

One must always be aware of disgruntled employees. Even acting alone, a knowledgeable individual could successfully carry out bioterrorism attacks. A better term for these attacks may be biocrimes.

Any biosecurity plan should provide for a well-established audit trail for specimens being collected and submitted for laboratory analysis. Life and property are of prime importance, but perpetrators must be prosecuted successfully.

An area that frequently gets overlooked is corpse and carcass disposal. In the case of a large incident, the task could be overwhelming and often exceeds everyone's imagination. The agent used may affect the manner in which disposal is accomplished.

One must always keep in mind that dissemination of agents by a bioterrorist will probably be far different than the expected large aerosols that would be encountered on the battlefield. One can also imagine that attempts by terrorists may be bungled due to lack of dissemination knowledge. Nevertheless, a few affected humans or animals may achieve the terrorist's goal of chaos and panic.

Following an incident, those in charge of responding must be ready to provide sound, understandable information to the public in an orderly, planned manner. Also they must be able to assure the public that they are safe or to provide safety measures if needed.

In the case of explosive or chemical incidents, police and fire personnel are considered the first responders. In the case of biological incidents, the first responders are likely to be personnel in emergency rooms and clinics; in the case of animals, veterinarians and farmers; and in the case of plants, farmers will probably be the first to recognize the problem.

Any strategy to develop nationwide biosecurity and response programs should be woven into programs addressing new and emerging diseases. Resources will be limited in all areas, but efforts that link such programs will benefit all programs.

Infecting Soft Targets

Biological Weapons and Fabian Forms of Indirect Grand Strategy

ROBERT D. HICKSON[a]

Department of Philosophy and Fine Arts, U.S. Air Force Academy, 2354 Fairchild Drive, Suite 6K12, U. S. Air Force Academy, Colorado 80840-6238, USA

Underlying the exposition of subtle deception and strategic indirect warfare that follows is the theme of trust, to include: the grave personal and cultural consequences of intimately broken trust and how the intimate effects of broken trust may themselves be strategically and grand-strategically manipulated by a deft opponent. The greatest social consequence of the lie is that it breaks trust. And trust, once broken, is so hard to repair, even with forgiveness, even with graceful mercy and the healing of the memory. Such a poignancy—such a fragility and vulnerability—is one of the unmistakable themes of all the world's great elegiac and tragic literature. It also pertains to the world of strategy and grand strategy, which also takes the longer view and goes to the roots of things.

As in a tragically fragmented family, a culture of broken trust, especially when it involves an intimately broken trust, is likewise self-sabotaging and often deeply destructive. Such a riven and wounded culture is thereby also more vulnerable to strategic exploitation and external maneuver by a subtle adversary. If, for example, an intelligent long-range adversary perceives the United States to be a "rogue superpower" and a "hectoring hegemon," but also a "declining hegemon" marked by a loss of purpose, decadence, and broken trust, he will likely also perceive how an exploitable weakness has favorably manifested itself, even as a "provocative weakness"—"so weak that it is provocative to others" (in the memorably accented words of Dr. Fritz Kraemer). When, moreover, increasingly untrustful American citizens are fearful of the safety of their food and their water, to include the long-range safety of genetically modified foods; and when the military culture itself is increasingly untrustful of the limited or experimental vaccines they are dubiously obliged to receive, others will likely notice our "internal contradictions" and "exploitable weaknesses," which all, at root, derive from a cumulative and innermost broken trust. Such adversaries, desiring to limit or to "level down" the United States, as well as Israel, for example, might well the "seize, retain, and exploit the initiative" strategically and grand-strategically, and thus further maneuver to subvert domestic trust.

Reality is that which doesn't go away, even when you stop thinking about it. If somebody is at war with you, even if you don't know it, you're at war! Furthermore, every assessment of a threat is correlative to the vulnerability of the target—to include the "target culture" and the target's vulnerable trust in its agriculture and sus-

---

[a]Address correspondence to: Dr. Robert Hickson, Department of Philosophy and Fine Arts, HQ USAFA/DFEG Hickson, 2354 Fairchild Drive, Suite 6K12, U.S. Air Force Academy, Colorado 80840-6238; Telephone: 719-333-8716; Fax: 719-333-7137.

tainable agricultural infrastructure. All strategy and responsive counterstrategy must first be attentive to the "security of its base," before it can also adequately achieve "mastery of the communications," which is itself a strategic indispensability, as well as a part of the maneuvering "preparation for the strategic advantage" (or what the Chinese call *shi'h*).

The use of biological weapons to infect food supplies, blood supplies, vaccines, water and other "soft targets" would constitute a formidable challenge to our nation and political culture, especially if it were also to be intelligently harnessed to Fabian forms of indirect grand strategy. This conjunction is a terrible thing to think upon, and yet we must do so, because history shows that indirect grand strategy, with its use of surprise, delay, and psychological dislocation, has been used repeatedly and effectively against militarily more powerful adversaries.

Fabian strategy is named for the Roman general Quintus Fabius Maximus (d. 203 BC), who defeated Hannibal by avoiding direct conflict. His long-range strategic indirection and evasiveness countered Hannibal's military genius and sapped the energy of his forces. (The Fabian Society, founded in nineteenth-century Britain, also adopted the strategy in an attempt to introduce socialism gradually and indirectly.)

If Fabian strategy were now to be used in intentionally incongruous and shocking combination with more immediately traumatic forms of bio-terrorism, this could rightly be understood as a new form of psychological or political warfare—for it specifically targets the human mind and the weakened or despairing will, especially of one's vacillating political leadership or fractious allies. For the purposes of this paper, I will discuss this form of warfare not just as an effective weapons system (albeit without conventional fire-power), but as an even larger new phenomenon that is more fittingly called strategic psycho-biological warfare, which exploits current revolutions in molecular biology and genetic engineering while aiming to manipulate the fears, broken trust, and uprooted hope of a modern citizenry at the end of a dark century.

Psycho-biological warfare, with its technical manipulations, ethical equivocations, and purposive confusions, could take us, finally, to the foundations of what it means to be a human person, as distinct from a mere artifact to be experimentally engineered and impersonally discarded. This could compel us, as well, to answer some trenchant questions: "What is a human person?" and "What is a human person for?" For how we see human life and its moral purposes[1] will profoundly affect the limits we set in warfare, especially in the fearsome and far-reaching realm of warfare considered here. Any adequate American grand strategy to counter psycho-biological warfare must first consider such moral limits; it must also consider the long-range aftermath of such warfare, which is so likely to stain the nature of the subsequent peace and have even deeper after-effects on civilization.

To appreciate these larger issues more fully, we must first turn to history and, specifically, to Israeli military history. When, in September 1949, the Chief of the General Staff of the Israeli Forces, General Yigael Yadin, wrote his stratregical analysis of the 1948–1949 Arab-Israeli War, he eloquently accentuated his understanding of and successful dependence on B. H. Liddell Hart's theory of indirect strategy, especially its psychological subtlety and efficacy.[2] What if, fifty years later, keen-minded anti-Israeli and anti-American strategic thinkers were to apply Liddell Hart's strategic principles against Israel and America? That is to say, what if adversaries now applied the insights of Liddell Hart in order to produce, both in Israel and in the United

States, strategic surprise, shock trauma, psychological dislocation, and paralysis, especially by manipulating the imagined or actual effects of bio-terrorism and longer-range biological warfare? The words of Liddell Hart should concentrate our attention: "It should be the aim of grand strategy to discover and pierce the Achilles' heel of the opposing government's power to make war."[3] In our own case, the aim could be to discover and pierce the Achilles' heel of the U.S. government's power to carry out what is sometimes perceived as its undefined, provocative, and increasingly resented "policy of engagement and enlargement" abroad. And indeed as with other great powers in history, the perception of our strategic policy as overbearing is likely to provoke "political jujitsu," as Saul Alinsky called it, and other Fabian forms of indirect grand strategy against us, is it not?

That is to say, strategic thinkers opposed to Israel and the United States may by now have "grasped what the soldier, by his very profession, is less ready to recognize—that the military weapon is but one of the means that serve the purposes of war; one out of the assortment which grand strategy can employ."[4] Once this larger and more inclusive understanding is grasped by an adversary, "the military principle of 'destroying the [enemy's] main armed forces on the battlefield'…fits into its proper place along with the other instruments of grand strategy—which include the more oblique kinds of military action as well as economic pressure [or economic warfare], propaganda, and diplomacy [or what General Beaufre, as we shall see, called the mentally dislocating 'exterior maneuver']."[5]

In this view,

> [i]nstead of giving excessive emphasis to one means,…it is wiser to choose and combine whichever are the most suitable, most penetrative, and most conservative of effort —i.e., which will subdue the opposing will at the lowest war-cost and minimum injury to the post-war prospect. For the most decisive victory is of no value if a nation be bled white in gaining it.[6]

Liddell Hart also proposed a complementary insight: "[T]his decisive strategic victory…was rendered indecisive on the higher strategic plane [i.e., of grand strategy]."[7] Even an effective indirect approach to the enemy's strategic rear, for example, may be nullified by a larger failure in grand strategy, to which lower, more physically decisive military strategy must always be subordinated, adds Liddell Hart:

> For, if the government has decided upon a limited aim or "Fabian" grand strategy [i.e., one of protracted indirection, delay, and evasion], the general who, even within his strategic sphere, seeks to overthrow the enemy's military power may do more harm than good to the government's war policy.[8]

In the Peloponnesian War between Sparta and Athens (431–404 BC), the Spartans initially had to face a kind of "Fabian" strategy and

> were foiled by Pericles's war policy, of refusing battle on land while using the superior Athenian navy to wear down the enemy's will by devastating raids. Although the phrase "Periclean strategy" is almost as familiar as the "Fabian strategy" in a later [Roman] age, such a phrase narrows and confuses the significance of the course that war pursued [after the 430 BC Plague in Athens]. Clear-cut nomenclature is essential to clear thought, and the term "strategy" is best confined to its literal meaning of "generalship"—the actual direction of military force, as distinct from the policy governing its employment and combining it with other weapons: economic, political, psychological. Such policy is in application a higher-level strategy, for which the term "grand-strategy" has been coined. In contrast to a strategy of indirect approach which seeks to dislocate the enemy's balance in order to produce a decision, the Periclean plan was a grand strategy with the aim of gradually draining the enemy's endurance in order to convince him that he could not gain a decision. Unluckily for Athens, the importation

of plague tipped the scales against her in this moral and economic attrition campaign. Hence in 426 BC, the Periclean strategy was made to give place to the direct offensive strategy of Cleon and Demosthenes.[9]

He also adds that "through the exasperation and fear that this [Spartan counter-offensive strategy] generated [i.e., "by taking an economic objective," the "Athenians' 'national' lines of communication"], he [the enemy Spartan general, Lysander] was able, thereby, also to produce conditions favorable to surprise and to obtain a swift military decision."[10] Later, ironically, the altogether weaker city-state of Thebes was able, gradually, to "[release] herself from Sparta's dominion by the method later christened Fabian, of refusing battle...."[11] Is it not also reasonable to suppose that the U.S.'s adversaries today might have similar incentives to resort to Periclean or Fabian indirection?

It is also important to consider that "the strategy of Fabius [known, interestingly, as the "Cunctator," or "Delayer"] was not merely an evasion of battle to gain time, but calculated for its effect on the morale of the enemy—and, still more, for its effect on their potential allies" and thus "was...primarily a matter of war-policy or grand strategy."[12] Says Liddell Hart:

> The key condition of the strategy by which this grand strategy was carried out was that the Roman army should keep always to the hills, so as to nullify Hannibal's decisive superiority in calvary. Thus this phase became a duel between the Hannibalic and the Fabian forms of the indirect approach.[13]

To what extent will the United States, as well as Israel, now have to face Periclean, Hannibalic, or Fabian forms of the indirect approach—and other insidious forms of "asymmetrical" indirection that use biological agents to achieve an even more devastating psychological effect of subversion and dislocation on the citizenry and soldiery? To what extent will biological warfare (and bio-terrorism) on our own home front now be—or be perceived to be—the U.S.'s "Achilles heel" and perhaps become an asymmetrical form of retribution for our obtrusive policy of "engagement and enlargement"? Given our current form of government and Constitutional provisions, how can we discern and counteract an adversary with biological weapons who also possesses strategic "interior lines" on the "inner front" of our homeland, so as to infect such vulnerable soft targets as vaccines, water, and food and blood supplies? A good strategist must first reliably secure his own base and become "master of the communications," especially the strategic lines of communication, both interior and exterior, the mass media, and the communications of his enemy. How will our defenses counter such subtle penetration?

We can gain insight into these questions from a noted French military strategist, General André Beaufre, writing in 1963 on indirect strategy and the psychological factor in war. His thoughts have trenchant implications for our situation in America today. Learning from the humiliations he had known both as a Frenchman and as a combatant commanding officer, he warned and instructed us about the insidious methods of indirect strategy.[14] America has much to learn from him.

Beaufre says that even though strategy can be played two ways, directly and indirectly—like the major and minor keys in music—the object of strategy remains the same: "a struggle for...freedom of action" leading to "a decision arrived at through the psychological surrender of the enemy," The object is "to produce a climax—the point at which the enemy's morale cracks." When, according to Beaufre, one is able "to strike terror, to paralyze, and to surprise" one's adversary—"and all these objects are psychological"—then one can limit or remove his freedom of action and his security, often by seizing, retaining, and exploiting the initiative and by "the strategy of guile." But, always in strategy, "the touchstone is freedom of action," especially when, as is usually the case, "psychological action must precede military action" and prepare for military action by a psychological "artillery barrage," which includes diversion and deception.

What is our own strategic freedom of action today in the United States, both psychologically and militarily, against foreign and domestic threats of bio-terrorism and longer-range psycho-biological warfare? How might our adversaries, at home and abroad, be preparing to distract and dislocate us, physically and psychologically? Since, according to Beaufre, strategy is "a thought process" and "the art of the dialectic of [at least] two opposing wills" in order to "reach the other's vitals by a preparatory process," how might U.S. strategists anticipate the use of biological weapons by such preparatory and insidious indirection?

Beaufre uses the forceful metaphor of "an incubator war," such as "the lethal but insidious infections of the Cold War or 'war in peacetime' (*la Paix-Guerre*)." He says that "in an incubator war, psychological infection [including the infection of panic] is not unlike that produced by biological warfare," for, "once launched, it is difficult to control," just as "the virus of Bolshevism rebounded upon her" after Germany enabled Lenin to arrive in 1917 at Finland Station in St. Petersburg to start his revolution. Beaufre proposes that the Soviet's revolutionary dialectic of dissolution against its enemies was, like biological warfare, "a method of slow creeping diffusion of chaos under the umbrella of an insidious threat." By using "psychological technology...partly camouflaged by an anesthetizing propaganda campaign," and by using Alinsky's "political jujitsu," the indirect strategy of the Soviets, says Beaufre, aimed "to disorganize the enemy by disrupting...[mental] cohesion...[and] loosening...moral ties." This strategic "enervation or erosion method," a part of the "new style of war," says Beaufre, is itself like "the creeping infection of an illness"—a gradual titration and permeation of an infection. Beaufre's metaphors are even more forceful when applied to the modern realm of psycho-biological warfare.

Against psycho-biological forms of warfare, as well as against new forms of Marxist or Gramscian revolutionary warfare (as seen for example in the Trans-National Radical Party in Europe today), there is a grave need for what Beaufre calls "inoculation and counter-infection," because they are part of a new battle for the mind. In the context of our vulnerable democratic culture, the challenges in forming an integrated defense-in-depth against psycho-biological warfare are great indeed.

In forming such a defense, it is important to note that even the best of tactics or operations (i.e., "the sum total of the dispositions and maneuvers") are "rendered nugatory," says Beaufre, "if used to further an erroneous strategy." Tactics "must be the servant of strategy," but the "choice of tactics is, in fact, strategy," as when deciding, for example, "whether to use force or subversion" as a subordinate part of one's own larger or grand "strategy of guile." As Beaufre says, "how total [i.e., how inclusive] the art of strategy must be"—because it involves politics, economics, finance, and psychology, among other things. "The strategic priority" must always be "to decide how great the freedom of action is for oneself and what is available to the enemy." In the face of biological weapons today, how would we ourselves make this decision?

We must also answer such questions as these: Who is the enemy? What (or whom) are we trying to protect? And why? The amount of access to our "interior lines" (i.e., to our interior dispositions, communications, and maneuver room) that is unwittingly provided to our adversaries, including trans-national criminal syndicates, is very great.

In my experience over the last three years, all too many people, when considering bio-terrorism and indirect biological warfare, have been cynically (or flippantly) inclined either to a kind of "pre-emptive futility" or to various forms of denial, both of which already constitute "pre-emptive psychological surrender"! "What can we do?" was the question put to me often enough. However, those whose special duty of leadership it is to provide for the common defense are called to a higher standard of foresight and determination.

It has been with these considerations of duty in mind that Dr. Thomas Frazier has worked so selflessly and indefatigably, despite discouragement and disincentives, to bring so many scientists, specialists, analysts, and thinkers together for candid discourse and a call to action. For merely passive forms of defense against psycho-biological weapons will likely be insufficient and perhaps even ruinous.

But, as to our response policy, should U.S. counter-initiatives be immediate and proportionate, like the well-known counter-initiatives of Israel? Would this be self-defeating for the U.S., exacerbating or only dissipating, given our diverse and vulnerable extensions abroad as well as our cultural politics at home?

One of the reasons, therefore, I am focusing our attention on Fabian forms of indirect grand strategy to make psychological use of bio-agents and bio-technologies is to make us more aware of the dangers of over-reaction, which might not only increase our vulnerability, but could even help unite additional hostile elements against us. That is to say, in the gathering disillusionment and resentment against the United States, many are likely to say "the enemy of my enemy is my friend." The consideration of Fabian indirection will also likely make us more vigilant to the rash and reductive "terrible simplifiers," those who might wish to use the pretext of a biological threat to implement Emergency Executive Orders or new "global arrangements" favorable to essentially unaccountable international oligarchies or NGOs (non-governmental organizations), but potentially unfavorable to national or local authority. I make this contention on the premise that a humane and proportionate scale—or scope of command—must be maintained when trying to deal with the threatened or actual conduct of psycho-biological warfare, because it specifically tries to destroy intimate trust, both in a community and in the individual mind.

In light of some fundamental axioms of strategy and grand strategy that will now be further elucidated by Liddell Hart, we will be able to consider more concretely how grand-strategic Fabian bio-warfare might operate and have its psychologically dislocating and paralyzing effects. Let us assume that an adversary or coalition of adversaries might wish to "revive the art and effect of strategy"[15]—especially long-range indirect grand strategy. The culture of China, for example, with its remarkable cultural cohesiveness over time and space, might be especially adept at grand-strategic deception. Certain European governments and Euro-socialists wishing for the diminishment of U.S. influence and enhancement of the euro as an international reserve currency might indirectly co-operate with China and others to add to America's discomfiture, by omission at least if not by commission. In the *London Mail*, for example, Allan Piper and Richard Grant write:

> The introduction of the Euro in January [1999] threatens to trigger the worst global economic crisis since the Second World War. It could even signal the breakdown of the global financial system, according to the City's [London's financial district's] most respected economist. Stephen Lewis, who provides daily advice to the Square Mile's leading institutions, blames the advent of the Euro for the present turmoil in world markets, and warns that massive currency movements created by its introduction will make matters worse. He predicts that, because European governments are *determined to break the power* of the U.S. dollar, it will encourage a worldwide proliferation of nationalistic policies, force widespread introduction of currency exchange controls, and lead to a sharp slowdown in global economic growth.... Lewis' remarks follow an announcement from Beijing last week that the Chinese government wants to offload dollars from its $140 billion foreign currency reserve to buy the Euro.... Lewis warns: "One of the reasons there is a crisis at all is that the governments sponsoring the Euro are *seeking to overturn the dollar's supremacy*. They do not want the dollar to survive as the world's leading currency. A large part of the global economic problem over the past year has arisen from attempts by policy-makers to assert the Euro's role in the scheme of things. This challenge is the biggest since 1945." Last week, Wang Jian, economist of China's State Development Planning commission, said that the country's government would cut the proportion of dollar holdings to 40% so that it could build Euro holdings to the same level.... He [Stephen Lewis] said: "The movement of capital will devalue the dollar sharply and cause economic recession in the U.S. The significant point about Wang's comment is that it came days after German bankers had been in Beijing seeking to persuade the authorities to shift their reserves from the dollar to the Euro." (Emphasis added.)[16]

In this context, additional disruptions from the use of actual or feigned bio-agents could be traumatic and dislocating. With this example in mind, Liddell Hart's axioms become even more cogent and sobering as we consider the Fabian use of biological weapons.

Liddell Hart is fundamentally opposed to two theses: (1) that "battle is the only means to the strategical end" and (2) that "in war every other consideration should be subordinated to the aim of fighting decisive battles."[17] He thinks it wise, instead, often "to enjoin a strategy of limited aim"[18] and especially "a limited aim or 'Fabian' grand strategy."[19]

He says:

> The more usual reason for adopting a strategy of limited aim is that of awaiting a change in the balance of force—a change often sought and achieved by draining the enemy's force, weakening him by pricks instead of risking blows. The essential condition of such a strategy is that the drain on him [e.g., the U.S.] should be disproportionately greater than on oneself. The object may be sought by raiding [or infecting] his supplies;...by luring him into unprofitable attacks [i.e., "lure and trap" or "mystify, mislead, surprise"]; by causing an excessively wide distribution [or centrifugal overextension] of his force; and, not least, by exhausting his moral and physical energy.[20]

When strategy, from its etymology, is considered as "generalship," it is "the art of distributing and applying military means to fulfill the ends of policy" (as well as the ends of grand strategy) by "the movement of forces" and "its effect," particularly when "the effect was [or will be] insidiously harmful."[21] The purpose of strategy, as well as grand strategy, is "to diminish the possibility of resistance" and "to fulfill this purpose by exploiting the elements of movement and surprise."[22] Says Liddell Hart:

> The role of grand strategy—higher strategy—is to co-ordinate and direct all the resources of a nation, or band of nations, towards the attainment of the political object...the goal defined by fundamental policy. Grand strategy should both calculate and develop the economic resources and man-power of nations.... Also the moral resources—for to foster a people's willing spirit is often as important as to possess the more

concrete forms of power.... Moreover, fighting power is but one of the instruments of grand strategy—which should take account of and apply the power of financial pressure, of diplomatic pressure, of commercial pressure, and, not least, of ethical pressure, to weaken the opponent's will.[23]

Even when it pertains to the lower level of strategy, Liddell Hart argues—and our new adversaries may have listened to him—that "strategy not only stops on the frontier [of the province of fighting], but has for its purpose the reduction of fighting to the slenderest possible proportions" and, if fighting is unavoidable, "to bring about battle under the most advantageous circumstances."[24] And sometimes, as in the case of the Greek Byzantine general, Belisarius, in Syria, "the national object" was fulfilled by "pure strategy," for, "in this case, the psychological action was so effective that the enemy surrendered his purpose without any physical action at all being required."[25] Liddell Hart comments:

While such bloodless victories have been exceptional, their rarity enhances rather than detracts from their value—as an indication of latent possibilities, in strategy and grand strategy. Despite many centuries' experience of war, we have hardly begun to explore the field of psychological warfare.[26]

With respect to the military strategist or grand strategist, Liddell Hart says, by way of summary:

His true aim is not so much to seek battle as to seek a strategic situation so advantageous that, if it does not of itself produce a decision, its continuation by battle is sure to achieve this.[27]

Now, with reference to those who would use biological weapons to effect "psychological action," we must remember that "dislocation is the aim of strategy" and the intended sequel is "the enemy's dissolution or his easier disruption in battle."[28] But "how is the strategic [or grand strategic] dislocation produced?"—by, for example, "a move directed towards the enemy's rear," "a menace to its [interior] line of communication," or seeking to gain "a decisive advantage previous to battle."[29] It may also be produced by "menacing [or ambushing] the enemy's [or the "first-responder's"] line of retreat," "menacing the equilibrium of his dispositions," or "menacing [or contaminating] his local supplies [including his medical supplies]."[30]

The proper strategic intention is not so much to produce strain, but rather to produce shock—suddenness and surprise. "Psychological dislocation fundamentally springs from the sense of being trapped."[31] Also, the "strategy of an indirect approach [is] calculated to dislocate the opponent's balance," physically or logistically but, especially, mentally. In fact, "paralyzing the enemy's action" is "what constitutes a strategic indirect approach," which is itself "preceded by distraction [i.e., "to draw asunder" the opponent], so as "to deprive the enemy of his freedom of action" and to give him the sense of being trapped. Such a preparatory distraction also seeks "the distention" and "the diversion" of the opponent's forces, with the result that they are "too widely distributed and committed elsewhere"[32] so as not to be able to regroup and effectively concentrate against one's own forces—that is to say, "not giving your opponent freedom [of action] and time to concentrate to meet your concentration."[33]

Given modern conditions and mobile weaponry, says Liddell Hart, "the need for [preparatory] distraction" has grown. The "most economic method of distraction" is to force on one's enemy a choice of disconcerting "alternate objectives" along a single line of operations—striving to constantly "[put] the enemy on the horns of a dilemma" (as Sherman did in his "deep strategic penetration" of Georgia).[34] Citing the two correlative principles of "concentration of strength against weakness" and "dispersion of the opponent's strength," Liddell Hart emphasizes that "true concentration is the fruit of calculated dispersion."[35]

Liddell Hart thinks it essential to "adjust your end to your means," after a sober assessment of one's means, and to "think what it is least probable that he [i.e., the enemy] will foresee and forestall."[36] Since "a single objective is usually futile," he says, it is important to "take a line of operations which offers alternative objectives." This is also "the basis of infiltration tactics,"[37] which today could include biological weapons, to exploit the opponent's confusion, mental dislocation, disorganization, and demoralization—and to exploit them before he or his society can recover. However, certain cautious and unstrategic minds, inordinately focused on tactics, tend to promote "the common indecisiveness of warfare," to "obscure the psychological element," and "to foster a cult of soundness rather than of surprise."[38]

One must bear in mind "the necessity of making the enemy do something wrong" and, "by compelling [his] mistakes," to "find in the *unexpected* the key to a decision."[39] For "a man unnerved is a highly infectious carrier of fear, capable of spreading an epidemic of panic."[40] Although strategy "should seek to penetrate a joint [or critical communications node] in the harness [or networks] of the opposing forces," Liddell Hart emphasizes that "a strategist should think in terms of paralysis, not killing."[41] But again, a "decisive strategic victory" can be "rendered indecisive on the higher strategic plane" of grand strategy.

Given the new face of terrorism, as seen for example in the Aum Shinrikyo cult, there is, it seems, a growing "fanaticism unmixed with acquisitiveness" and "infused with the courage of desperation."[42] This new enemy seeks only to destroy, not to conquer —and biological weapons will serve him well.

By taking the measure, in the larger grand-strategic context, of both the capacities of biological weapons today (actual and potential) and the resentful intentions of terrorists or transnational criminal syndicates, our judgments and responses will be more disciplined and wiser, more prudent and proportioned. We must not think of biological weapons or bio-terrorism in merely tactical or operational terms, or in isolation. We must anticipate and consider them in the context of Fabian forms of indirect grand strategy, which may subtly employ new biotechnologies and discoveries from neuroscience, such as psychotropic and neurotropic bioagents, to infect the human mind and weakened will. Such subtle forms of strategic indirection against "soft targets" aim to subvert trust, the most intimate forms of trust, thereby producing, if not our despair and desolation, then, at least, our demoralization and strategic paralysis.

REFERENCES

1. HARRIS, J., 1992. *In* Wonderwoman and Superman: The Ethics of Human Biotechnology. P. Singer, Ed. Oxford University Press. New York.
2. LIDDELL HART, B.H. 1967. Strategy. 2nd edit. Meridian Books. New York.
3. Ibid. p. 212.
4. Ibid. pp. 211–212.
5. Ibid. pp. 211–212.
6. Ibid. p. 212.
7. Ibid. p. 237.

8. Ibid. p. 321.
9. Ibid. p. 10.
10. Ibid. p. 13.
11. Ibid. pp. 13–14.
12. Ibid. p. 26.
13. Ibid. p. 27
14. BEAUFRE, A. 1965. An Introduction to Strategy, Praeger. New York. Quoted text from pages 1, 23–24, 30, 34–35, 42, 47, 55–57, 59, 80, 83, 86, 99, 100, 102–104, 108–110, 121–122, 127–128, 133, 135, 137–138.
15. LIDDELL HART, op. cit. p. 332.
16. PIPER, A. & R. GRANT.1998. London Mail (6 Sept.): 1.
17. LIDDELL HART, op. cit. p. 319.
18. Ibid. p. 320.
19. Ibid. p. 321.
20. Ibid.
21. Ibid. pp. 321, 319.
22. Ibid. p. 323.
23. Ibid. p. 322.
24. Ibid. p. 324.
25. Ibid. p. 325.
26. Ibid.
27. Ibid.
28. Ibid.
29. Ibid. p. 326.
30. Ibid.
31. Ibid.
32. Ibid. p. 328.
33. Ibid. p. 334.
34. Ibid. p. 339.
35. Ibid. p. 334.
36. Ibid. p. 335.
37. Ibid.
38. Ibid. p. 336.
39. Ibid. p. 336 (Emphasis added).
40. Ibid. p. 212.
41. Ibid.
42. Ibid.
43. Ibid. p. 359.

The First Step toward Building Tools and Methods for Protection

JOSEPH D. DOUGLASS, JR.[a]

Consultant, Falls Church, Virginia 22046, USA

Playing catch-up ball can be exceedingly difficult, especially when you are in an area that is changing rapidly, as has been the case with chemical and biological warfare (CBW) technology in the last thirty years.

In 1969–1970, President Nixon unilaterally renounced biological and toxin warfare and said the United States would never be the first to use chemical weapons. Biological, toxin, and chemical warfare arms control treaties were ratified by Congress in 1973. In effect, the United States unilaterally took itself out of the CBW ball game. Military and intelligence service efforts dropped to almost zero almost overnight. This decrease was greatly facilitated by the fact that no one likes CBW, especially the military services. A contaminated battlefield is very difficult to cope with; it is hard to survive; and it is close to impossible to fight in such environments. The services were only too pleased to have an excuse to stop worrying about CBW.

In retrospect, the decision of "opting out" could not have come at a worse time. The Soviets, then our main enemy, were expanding their efforts in CBW and the underlying technology base was entering a revolutionary phase.

SOVIET EXPANSION

The Soviet expansion in chemical and biological warfare had begun in 1965. In 1966, they tested battlefield concepts with live agent employments. In 1967, plans for conducting CBW in the event of a war with NATO were approved and formally incorporated into Warsaw Pact plans. Release and employment was predelegated—to division commanders for chemical warfare and to the Warsaw Pact commander for biological warfare. Research and development also accelerated in 1967 with the adoption of an aggressive 20-year plan, whose goal was the development of qualitatively new families of chemical and biological weapons. This was an extremely important decision. The Soviets believed that future chemical and biological weapons would surpass nuclear weapons in importance. They believed that nuclear weapon stockpiles would be greatly reduced through arms control efforts and that advanced chemical and biological weapons would make them even stronger *vis a vis* the United States when nuclear stockpiles were reduced.

Preparation for this development of new chemical and biological weapons began in the mid 1960s with initial focus on training scientists and technicians. Massive

[a]Address correspondence to: Joseph D. Douglass, Jr., Ph.D., Consultant, 203 Garden Court, Falls Church, Virginia 22046; Telephone: (703)533-9452; Fax (703)532-8685.
e-mail: red@wt.infi.net

construction of biological warfare facilities actually began with the approval of the new five-year plan in 1972. This was the start of the program that would become known as Biopreparat. By 1990, the research, development, and production for just biological warfare, as described by Dr. Ken Alibek, involved more than 50,000 people—more than 32,000 in Biopreparat, more than 10,000 in Ministry of Defense biolabs, more than 10,000 in the Ministry of Agriculture, and thousands more in various agencies and research facilities of the Academy of Sciences. Biopreparat alone involved more than 40 facilities, some of which contained more than 100 buildings. While this is enormous when compared with our perception of the Soviet effort in the late 1980s (before the sequence of defectors from the Biopreparat program), it still represents, in my judgement, less than 25% of the problem, which involves much more than the defectors of note appear to have had access to.

THE REVOLUTION IN BIOTECHNOLOGY

While these Soviet developments went largely unrecognized for twenty years, there was no surprise concerning the second aspect, the parallel massive expansion in the underlying technology base. This has been called the biotechnology revolution. It received major attention, even from Wall Street, as new companies were formed to exploit the revolutionary promises, especially in the late 1970s and early 1980s. This technology base included a wide variety of technical advances that have become almost household words—for example, molecular biology, biochemistry, genetic engineering, gene splicing, monoclonal antibodies, and microencapsulation.

It was just as this technology was starting its meteoric rise and the Soviets were organizing to exploit it with unlimited resources that the United States military and intelligence services quit the field. The few individuals who remained in the area found it difficult to counter the nerve agent threat of the 1960s, much less to consider what the massive Soviet research and development programs would produce. This 1960s nerve agent threat became even more of a problem following the capture of Soviet equipment during the 1973 Middle East War. Intelligence experts found that this equipment, much to everyone's surprise and consternation, was fully equipped to operate in contaminated environments. In response, there were major efforts to play catch-up ball beginning in the late 1970s. Even this was an uphill battle because it was still hard to get attention focused in areas that the services really did not want to consider. Moreover, there was a great reluctance to think beyond the problems of sarin, soman, and VX because there was no intelligence and because the problems posed by sarin, soman, and VX were already "too hard to handle."

Worse still was the biological warfare situation because any recognition that there might be a biological warfare threat was tantamount to saying the arms control treaties were being violated, which no one wanted to admit was the case. It was not that we did not know something was happening. There was plenty of evidence. There was massive construction that was recognized by 1975. Top-level U.S. National Security Staff officials asserted that this was unimportant and that there was no real strategic value in biological warfare and, therefore, the Soviets would not be violating the treaty. Notwithstanding these official pronouncements, evidence continued to accumulate. There was evidence of human experiments, and then of the use of toxins and biological agents in Afghanistan and Southeast Asia. Also, a large explosion at Sverdlovsk released quantities of a biological warfare agent that killed a number of people who were downwind of the explosion. All these events were evidence of the blatant Soviet violation of the arms control treaties. One or two people worked hard to focus attention on these events—but it was like swimming against the tide. Numerous arms control advocates and scientists worked hard to discredit the intelligence as the few insiders worked to bring it to people's attention—those trying to discredit the intelligence had more money, more press, and the full support of the arms control community. Equally important, they were on the side of the longstanding national policy that directed people in government to avoid doing anything that would embarrass the Russians. Accordingly, efforts to focus attention on the modern technology side of the CBW threat were stillborn.

To fully appreciate the situation, compare the efforts underway in the 1970s and 1980s. The Soviets had massive research and development programs in both areas, with more than 50,000 people working on biological warfare, as indicated earlier. In addition, we may assume a similarly impressive effort on the chemical warfare side. In contrast, the United States had no comparable advanced technology in chemical or biological warfare research, development, or production and very little intelligence. Most intelligence was directed at understanding the 1960s nerve agent threats and associated equipment performance, and even those efforts were minimal. In terms of those focused on trying to understand where the Soviets were headed—that is, in understanding their advanced technology chemical and biological warfare development efforts—you could count the numbers of people involved on the fingers of your hands. For all practical purposes, the United States had totally withdrawn from the field.

The problem with playing catch-up ball after you have been out of the game for a long time—in the chemical and biological field, at least twenty years—is that you lose the base from which to recover. There is no short-term solution and the usual approach of "throwing lots of money at the problem" is unlikely to work because the understanding required to know what to do with the money is simply not there. It takes time to build that "critical mass" of understanding and expertise required to manage the money and achieve something effective in the process. This "understanding" has two sides to it. First, there is the technology, which has experienced truly massive growth in the past thirty years. Second, there is the understanding of the operational implications; that is, what can be done with the new technology. In the case of CBW, this is especially important because of the Soviet objective, which was the creation of *qualitatively* new families of chemical and biological weapons; that is, of new effects, new concepts, and new applications. This type of thinking—a significant break with tradition—had also been absent from the national security and intelligence community thinking since 1969.

Today, the problems are additionally difficult because there are a variety of related areas with which we are concerned and which have been difficult to deal with because of the national policy not to say anything or think anything that would reflect badly on the Russians or Chinese. These policies go back a long way. For example, as explained in Congressional hearings two years ago by Col. Phil Corso, who was Eisenhower's intelligence aide in the 1950s, the policy of not embarrassing the Soviets was clearly stated and it constrained actions: "We called this the 'Fig Leaf

Policy.' It paralyzed every effort against the Soviet Union...we shouldn't put pressure on the Soviet Union, we shouldn't try to detach a satellite.... We shouldn't be strident against the Soviets. That was policy that was written...they had our prisoners and we couldn't put pressure on them. That was it. Our policy forbid us from doing it. If you did it, you were disobeying national policy." Joe Goulden recognized the same policy in operation at the field command level: "The division's public information officer had killed [messages from President Rhee to the Republic of Korea's embassy in Washington] because some of Rhee's messages to other chiefs of state criticized the Soviet Union, 'contrary to command policy and the express order of the 24th Division.'"[1]

In the early 1970s, intelligence on Chinese narcotics trafficking, which was being employed as a type of chemical warfare, was suppressed by the White House so as to avoid interfering with the new initiative toward China. As General Lew Walt, who headed a commission that was sent to Vietnam to learn where all the drugs were coming from, said later, "Keeping silent about the Chinese drug trafficking was the most damnable order I ever received." As observed by J. H. Turnbull in his study of the role of Chinese drug trafficking: "The covert dissemination of opium narcotics for commercial and subversive purposes represents on of the gravest threats to the armed services and societies of the Free World.... The subversive operations must be recognized as a peculiar form of clandestine chemical warfare, in which the victim voluntarily exposes himself to chemical attack."[2]

Particularly serious, from our perspective today, was the tremendous effort to kill any recognition of the Soviet effort to organize, train, support, and finance international terrorism that was mounted from within the CIA in the early 1980s. During the development of a special intelligence study on terrorism, various elements and key individuals within CIA worked hard to avoid any conclusions that held the Soviets responsible. They went so far as to discredit their own sources in an effort to turn aside a detailed Defense Intelligence Agency (DIA) study of Soviet involvement in international terrorism. They would have succeeded, except that representatives from the National Security Agency (NSA) refused to discredit their sources, with the result that the DIA analysis was allowed to stand. Nevertheless, the Soviet efforts in this area, and other connected intelligence areas, were downplayed and never received the careful scrutiny and understanding they should have received.

This is critical because CBW was an important part of Soviet sabotage plans and training, which were closely related to terrorist plans and training. As a result, there are very large gaps in our knowledge of Soviet and Chinese operations that have important bearing on terrorism and sabotage capabilities today. That is, CBW is not the only area in which we are playing catch-up ball. The potential problem this creates is unmistakable when we recognize the tremendous number of trained intelligence operatives, not just in Russian intelligence today, but those who have "left the service" or those thousands from Eastern Europe who were left with no home or occupation when their own intelligence services suddenly disintegrated. When you examine the explosive growth in various nefarious activities immediately following the breakup in the Soviet Union—organized crime, narcotics trafficking, and industrial espionage—it is easy to conclude that what we are seeing is not mere evolutionary growth of these activities, but rather the rapid movement of serious and well-trained professionals into these activities.

This is a rather lengthy introduction to my concern over the tools and methods we now need to achieve a modicum of protection. Certainly, money and high level attention are needed. However, just throwing money at the problem can be counterproductive because it will generate so much action and confusion that it will take on a life of its own—a life that, by definition, has to be guided by hundreds of people who have wonderfully sincere objectives, but very limited understanding of the problem. More often than not, they don't know "where the bodies are buried," and where and why knowledge is absent and what is involved in filling those gaps. This is not meant in a degrading sense. It only recognizes the consequences of not being in the ballgame for well over twenty years. After twenty years, there is not much corporate wisdom or memory remaining. I hope this is all a figment of my imagination and is an inaccurate description of the current situation. After all, we have now had two or three years to catch up. Let me say I am merely expressing some concerns I have based on seeing similar efforts in the past in a number of different areas. Accordingly, it really seems to me that before we can move on in an effective manner, we need to pause and learn what we have missed in the past thirty years that is relevant to the current problem.

As an example of what I mean relative to the problem of biological warfare and agricultural products, there were major plans for using contaminated agricultural products as a means of disseminating chemical and biological warfare agents in preparation for war. I have never seen anything to suggest that consideration of these plans ever made it into various threat assessments. Additionally, I would also be very concerned about what an organization like the Aum Shinryn cult—but one that is technically and operationally competent—might do to attack major U.S. agricultural products. There is little question in my mind but that such plans were prepared and are simply waiting in the wings for the correct opportunity. These are only two of a very large number of possibilities, and realities, that come to my mind, but which are politically incorrect to imagine because of the politics that go back several decades.

Unfortunately, these problems remain ever present. We have recently witnessed an excellent example. I refer to the Rumsfeld bipartisan review of the missile threat to the United States whose results were announced this past July. In effect, they trashed the CIA study that placed the threat 15 years in the future, mainly because of the way in which the questions put to the CIA were asked. The study served political interests, but whether or not it served the United States and our planning for the common defense is a much different question.

The problems facing those who are concerned about chemical and biological warfare and terrorism or sabotage are very similar. The first step toward developing the tools and methods for countering these CBW threats is not to simply throw money at them, but, rather, to invest the time needed to understand them better. This is not easy because it will require a considerable amount of patience to fill the many gaps that were left by close to three decades, and because there are still people who do not want those gaps exposed or filled. That is, what I am suggesting is that there are four different aspects to the threat problem. The easiest part—easiest, not easy—is the technology. The more difficult aspects are the political, institutional, intelligence and counter-intelligence. It would be a gross error, in my judgement to focus more attention on the technical side than is focused on these other aspects.

REFERENCES

1. GOULDEN, J. 1982. Korea: The Untold Story. McGraw-Hill. New York.
2. TURNBULL, J. H. 1972. Chinese Opium Narcotics: A threat to the survival of the West. Foreign Affairs Publishing Co. Richmond, Surrey, England.

Targeted Immune Design Using RNA Immunization

ROBERT R. GARRITY[a]

Biological Mimetics, Inc., Frederick, Maryland 21701, USA

Extreme remedies are very appropriate for extreme diseases.

HIPPOCRATES, 400 B.C.

INTRODUCTION

Infectious disease is the unwanted partner of humankind. In today's world, old and emerging diseases are appearing on two insidious fronts. One front is via naturally occurring infection, including an infamous list of bacterial, fungal, viral, and parasitic pathogens that have plagued mankind throughout history. The second front is the product of humankind's dark obsession to use infectious agents as biological weapons: smallpox, anthrax, bubonic plague, botulinum toxins, ricin, and staph enterotoxins headline a growing list of human bioweapons. Others, such as Newcastle disease virus, foot-and-mouth disease virus, rinderpest, and African swine fever virus are becoming recognized as potential bio-threats to the food industry as well. It is clear that most, if not all, infectious agents can be adapted and used as potential biological weapons. They all pose significant threats and require rapid and effective countermeasures of surveillance and protection.

This may be especially true for viral influenza. There have been four major pandemics between 1918 and 1977. All have occurred in approximately 10- to 20-year intervals. The Spanish flu pandemic of 1918 resulted in the global death of more than 20 million people. Evidence suggests that most pandemic and major epidemic influenza viruses originate in Southeast Asia. This has led to an elaborate worldwide infrastructure to identify strains and prepare a trivalent vaccine against each annual threat. Isolation, scale up, and strain inactivation follow as each new year's whole-inactivated vaccine is prepared. The yearly expense of supporting this worldwide infrastructure, not to mention the man-hours lost to disease, is staggering. The costs are in the tens of billions of dollars.

Influenza viruses are divided into types A, B, and C, plus subtypes of A. Types A and B mutate constantly and result in the need for a new vaccine against different strains each year. Each annual vaccine contains two currently circulating strains of influenza A and one of influenza B. The problem with making a broad-based vaccine is complicated by the subtype diversity inherent in influenza A. There are 15 known subtypes of influenza A virus, all of which exist in the aquatic bird reservoir.[1] In addition to aquatic birds and humans, influenza A infects other warm-blooded animals

---

[a]Address correspondence to: Dr. Robert R. Garrity, Chairman and CEO, Biological Mimetics, Inc., 431 Aviation Way, Frederick, Maryland 21701.

including sea mammals, minks, chickens, turkeys, horses, and swine. Genetic and biological observations suggest that swine may serve as intermediates for the direct transfer of influenza A virus into humans.[2]

Influenza viruses are subtyped based on two viral surface proteins, hemagglutinin (HA) and neuraminidase (NA). The antibody to HA prevents subtype-specific infection. Neuraminidase probably functions to breakdown mucin in the respiratory tract, facilitating viral infection. Mutation through antigenic drift and genetic reassortment from one host to the next through antigenic shift in both HA and NA lead to antigenic diversity and the evolution of newly pathogenic strains. The Asian influenza virus of 1957 and the Hong Kong influenza virus of 1968 were pandemic strains of human influenza virus that contained a reassortment of genes from human and avian strains.

Until recently, only HA subtypes H1–3, and NA subtypes N1 and N2 were believed to exist in man. In May of 1997, Hong Kong officials reported the death of a three-year-old boy from complications of influenza. Shortly after his death, an H5N1 strain of influenza was isolated from the boy. Avian influenza, the "bird flu," is an H5N1 strain of influenza A recognized to cause infection in shore birds and chickens. This was the first time this particular strain was found in humans. Within months, 18 confirmed human cases of H5N1 influenza virus infection were identified. Six cases resulted in death. Faced with this public health threat, officials began the massive slaughter of 1.4 million chickens and other poultry in Hong Kong. Similar outbreaks in Pennsylvania in 1983 and 1997 and in Mexico in 1995 led to the mass slaughter of millions of chickens. Concerns that new H5 or H7 outbreaks in poultry could lead to highly pathogenic strains threaten not only the food industry, but threaten the overall state of public health worldwide. This potential threat requires sophisticated measures of protection and surveillance.

This paper briefly discusses a combination of methods that we have developed: (1) a targeted immune design by immune refocusing, designed to overcome antigenic diversity and (2) *in vivo* gene delivery by direct RNA immunization, designed to genetically deliver native immunogens. Both can be coupled to generate highly specific and broadly cross-immune responses for use as a vaccine, immunodiagnostic, or therapeutic.

IMMUNE EVASION

Immune evasion is a global strategy used by all pathogens. All pathogens are challenged to survive in a very hostile immune environment. Yet only 25 pathogens have been targeted by licensed human vaccines. There are far more vaccine-resistant pathogens for which no effect vaccine has ever been made. All vaccine-resistant pathogens have evolved exquisite evasion strategies to avoid immune clearance. Many use immunodominant decoy epitopes, or decoptopes as they have become to be known, as a strategy to confuse and dysregulate host immunity. These epitopes are characteristically charged, loaded with immune targets, and hypervariable. They serve as beacons for early immune effector functions, inducing robust B- and/or T-cell responses. However, it appears that many of these responses have little or nothing to do with protection. Indeed, many of these responses lead to immune paralysis and dysregulation. Early dominant pathogen-specific immune responses have been shown to limit polyclonality to multiple epitopes; limit maturation of effector and memory responses; cause immune dysregulation via cytokine imbalances and superantigenic induction of T- and B-cells; induce pathogenic-enhancing responses; dysregulate through autoimmune induction; induce blocking antibodies and cell-mediated inhibitors; cause immunosuppression; and alter programmed cell death. Paradoxically, their success could also be their Achilles heel. Removal of these immunodominant dysregulating responses can be used to overcome pathogen-induced immune circuitry that insures pathogenic survival and continued transmission. One of many examples of this is illustrated in leishmaniasis.

Human leishmaniasis is endemic in the Middle East. Infection by leishmania can lead to debilitating lesions in man. Immunological infection of leishmania major in mice had led to a better understanding of how the antibody- and cell-mediated arms of the immune system work. For simplicity, the antibody arm clears cell-free pathogens, and the cell-associated arm clears cell-associated pathogens. Induction of one immune arm regulates induction or suppression of the other. An immune response to a specific pathogen like leishmania results in a finely regulated balance between each arm. In lethal murine challenge models BALB/C mice are genetically susceptible and die shortly after infection. Natural infection induces an early immunodominant response to the so-called LACK antigen, which has been mapped to a 16-amino-acid stretch.[3] This results in an early, robust antibody response controlled by T-cells of the Th2 phenotype. An early burst of IL-4, a cytokine that instructs Th2 development, has been shown to be responsible for leishmania susceptibility.[4] In the absence of IL-4, the animals default to a Th1 cell-mediated pathway. In leishmaniasis, the cell-mediated pathway facilitates clearance of the disease. Transgenic mice can be derived that are unable to mount an immune response to LACK.[5] These LACK-tolerant mice genetically switch over to a Th1 response upon infection and are totally protected when infected with a lethal leishmania challenge. Interestingly, in susceptible animals, an early immune response to LACK has been linked to the cross-reactive recall of another innate response. Thus, susceptible animals have an innate immune memory for LACK. Once induced, this memory response rapidly leads to Th2 induction, subsequent Th1 suppression, susceptibility to disease, and propagation of the parasite. A protective Th1 immune circuit can be wired if the immunodominant LACK immune response is suppressed.

We have demonstrated that a new hierarchy of more broadly neutralizing immune responses can be induced to the major HIV-1 envelope glycoprotein through targeted immune dampening.[6] Establishing these responses up front, as part of an immune memory arsenal, is a potential way to defeat many pathogens that have resisted previous vaccine approaches.

TARGETED IMMUNE DESIGN

One of the major obstacles in generating a protective vaccine or broad-based therapeutic against influenza is to overcome the problems associated with antigenic diversity and immune dysregulation. We have developed a technology to refocus the immune response away from immune dysregulating epitopes. The HA antigen, the major target for neutralizing antibody, has been crystallized. This structure has corroborated the external presence of five previously identified hypervariable antigenic sites involved with viral escape and the evolution of new epidemic strains.[7] It is believed that restriction of the B- and T-cell repertoire to limited immunodominant and cross-reactive epitopes of the HA molecule during primary viral infection is responsible for this effect. Primary exposure to influenza results in establishment of a cross-reactive non-protective HA recall that is restimulated upon subsequent infection by antigenically distinct strains. This phenomenon is known as original antigenic sin (OAS).[8] OAS results in restricting the maturation of a more diverse immune response, a strategy used by influenza to insure superinfection in subsequent years by variant strains. Immune refocusing away from these immunodominant, dysregulating HA epitopes could break the dysregulating recall circuit and make a single vaccine or highly conserved antibody response to all strains possible.

RNA-MEDIATED GENE DELIVERY

It has been known for some time that phenol treatment of cells or tissues infected with a number of RNA viruses resulted in an aqueous fraction that was infectious.[9–11] Infectivity of this fraction was lost in the presence of ribonuclease, which had no effect on inactivation of intact virus. These are some of the earliest reports that direct intramuscular injection of genomic RNA could result in protein expression *in vivo*. Later studies showed that in addition to plasmid DNA, *in vitro*–generated RNA could also be injected directly into mouse skeletal muscle and result in reporter gene expression.[12] Taken collectively, these studies gave rise to the intriguing possibility that direct inoculation of genetic RNA had the potential to induce an immune response to the expressed antigen. Intramuscular injection of *in vitro*–generated RNA carrying self-replicating signals from Semliki Forest virus induced both antibody and cytotoxic T lymphocytes (CTL) responses to expressed antigens.[13] Direct inoculation of RNA encoding carcinoembryonic antigen (CEA) was shown to be an effective priming agent as subsequent boosting with cells expressing CEA raised higher anti-CEA responses than in unprimed animals.[14] We have demonstrated that direct intramuscular injection of *in vitro*–generated gp120 and gp160 runoff transcripts formulated in RNase-free water results in the induction of an anti-gp120 antibody response.[15] This is especially important in the case of HIV-1 gp160 immunization. Gp160 expression requires the co-induction of the regulatory protein Rev. Rev facilitates transfer of gp160 RNA out of the nucleus into the cytoplasm where the RNA is translated into protein. In the absence of Rev, the RNA remains in the nucleus and never gets translated. In addition, co-expression of Rev and gp160 is complicated by the complexities of alternate gene splicing and low expression of gp160. Gp160 RNA can be directly inoculated into muscle in the absence of Rev. The transcript finds its way into the cytoplasm of a host cell and is translated into protein directly in the cytoplasm obviating the need for any Rev-mediated nuclear transport.

In addition, a number of safety and regulatory considerations currently under evaluation concerning the negative effects DNA inoculation may have on the host make RNA immunization a useful alternative. Anti-DNA antibodies are associated with clinical features of systemic lupus erythematosus, which heightens concern as to whether an antibody response to inoculated DNA will manifest the same results.

In addition, gene expression and the presence of plasmid DNA have been reported to occur for up to 19 months following DNA-mediated inoculation.[16] Autoimmunity, tolerance, hypersensitivity, and immune complexing are all hypothetical outcomes of chronic antigenic expression. Long-term expression of tumor cell antigens in certain cancer vaccine strategies may potentiate oncogenic transformation. The foremost concern, however, with respect to oncogenic potential of *in vivo* DNA inoculation is the integration of vector DNA into chromosomal DNA. This raises the possibility that a neoplastic triggering event occurs through protooncogenic induction or tumor suppressor inactivation. As a result, the Center for Biologics Evaluation and Research recommends that clinical protocols evaluating DNA vaccination include assays for the possibility of DNA integration.[17] All of these concerns are obviated or significantly reduced using RNA immunization. Although it remains to be seen whether RNA lability will preclude its use, RNA immunization may be the simplest method to date to raise a conformationally faithful immune response.

CONCLUSION

In 400 B.C. Hippocrates noted that "Extreme remedies are very appropriate for extreme diseases." This Hippocratic aphorism is reasonably consistent with the overall theme of this volume on food and agriculture security. That is, in today's world we are unfortunately sensitized to extreme threats that many believe have an imminent potential. These threats require innovative methods of surveillance, response, and prevention. We have developed a technology to focus immune responses to conserved targets and genetically deliver an antigen through RNA immunization. Hopefully, this begins to complement the challenge of surveillance and add one or more protective vaccines to the list. Together, through the combined and diverse group of expertise brought together in this conference, we can begin to meet these extreme and emerging challenges on many fronts.

REFERENCES

1. WEBSTER, R.G. 1998. Influenza: An emerging disease. Emerg. Infect. Dis. **4:** 1–7.
2. Molecular basis for the generation in pigs of influenza A viruses with pandemic potential. J. Virol. **72:** 9367–7373.
3. MOUGNEAU, E., A. FEDERIC, A. WAKIL, S. ZHENG, T. COPPOLA, Z-E. WANG, R. WALDMANN, R. LOCKSLEY & N. GLAICHENHAUS. 1995. Expression cloning of a protective leishmania antigen. Science **268:** 553–556.
4. LAUNOIS, P., I. MAILLARD, S. PINGEL, K. SWIHART, I. XEENARIOS, H. ACHA-ORBEA, H. DIGGELMANN, R. LOCKSLEY, H. MACDONALD & J. LOUIS. 1997. IL-4 rapidly produced by Vβ4 Vα8CD4+ T cells instructs Th2 development and susceptibility to leishmania major in BALB/c mice. Immunity **6:** 541–549.
5. JULIA, V., M. RASSOULZADEGAN & N. GLAICHENHAUS. 1996. Resistance to Leishmania major induced by tolerance to a single antigen. Science **274:** 421–423.
6. GARRITY, R.R., G. RIMMELZWAAN, A. MINASSIAN, W.-P. TSAI, G. LIN, J.-J. DE JONG, J.GOUDSMIT & P.L. NARA. 1997. Refocusing neutralizing antibody response by targeted dampening of an immunodominant epitope. J. Immunol. **159:** 279–289.
7. WILEY, D. & J. SKEHEL. 1987. The structure and function of the hemagglutinin membrane glycoproteins of influenza virus. Annu. Rev. Biochem. **56:** 365–394.
8. FRANCIS, T., JR. 1953. Influenza: New acquaintance. Ann. Intern. Med. **39:** 203–221.

9. GIERER, A. & G. SCHRAMM. 1956. Infectivity of ribonucleic acid from tobacco mosaic virus. Nature **177:** 702–703.
10. COLTER, J.S., H.H. BIRD, A.W. MOYER & R.A. BROWN. 1957. Infectivity of ribonucleic acid isolated from virus-infected tissues. Virology **4:** 522–532.
11. BROWN, F. & D.L. STEWART. 1959. Studies with infective ribonucleic acid from tissues and cell cultures infected with the virus of foot-and-mouth disease. Virology **7:** 408–418.
12. WOLFF, J.A., R.W. MALONE, P. WILLIAMS, W. CHONG, G. ACSADI, A. JANI & P.L. FELGNER. 1990. Direct gene transfer into mouse muscle *in vivo*. Science **247:** 1465–1468.
13. ZHOU, X., P. BERGLUND, G. RHODES, S.E. PARKER, M. JONDAL & P. LILJESTROM. 1994. Self-replicating Semliki Forest virus RNA as recombinant vaccine. Vaccine **12:** 1510–1513.
14. CONRY, R.M., A.F. LOBUGLIO, M. WRIGHT, L. SUMEREL, M.J. PIKE, F. JOHANNING, R. BENJAMIN, D. LU & D.T. CURIEL. 1995. Characterization of a messenger RNA polynucleotide vaccine vector. Cancer Res. **55:** 1397–1400.
15. GARRITY, R.R., P.L. NARA, G. LIN & S. JOHNSON. 1997. Genetic vaccination with naked HIV-1 RNA results in low-titer anti-gp120 response. *In* Vaccines 97. Molecular Approaches to the Control of Infectious Diseases: 151–156.
16. WOLFF, J.A., J.J. LUDTKE, G. ACSADI, P. WILLIAMS & A. JANI. 1992. Long-term persistence of plasmid DNA and foreign gene expression in mouse muscle. Hum. Mol. Genet. **1:** 363–369.
17. ZOON, K.C. 1996. Points to consider on plasmid DNA vaccines for preventive infectious disease indications. FDA Points to Consider. Center for Biologics Evaluation and Research. Rockville, MD.

Sensitive and Rapid Identification of Biological Threat Agents

J. A. HIGGINS,[a,c] M.S. IBRAHIM,[b] F. K. KNAUERT,[b] G. V. LUDWIG,[b] T. M. KIJEK,[b] J. W. EZZELL,[b] B. C. COURTNEY,[b] AND E. A. HENCHAL[b]

[a]*U.S. Department of Agriculture–Agricultural Research Service, Beltsville, Maryland 20705, USA*

[b]*Diagnostic Systems Division, U.S. Army Medical Research Institute of Infectious Diseases, Fort Detrick, Maryland 21702, USA*

INTRODUCTION

The threat of biological warfare and bioterrorism has increased in the last two decades. Classical biological threats that were part of the now-defunct U.S. offensive program (terminated in 1969) included *Bacillus anthracis*, Botulinum toxin, *Francisella tularensis*, *Coxiella burnetii*, Venezuelan equine encephalitis (VEE), *Brucella suis* (brucellosis), Staphylococcal enterotoxin B (SEB), rice blast, rye stem rust, and wheat stem rust. Most of these agents can be produced cheaply by aggressors and could have a larger impact on both military and civilian populations than other weapons of mass destruction.[1] In the hands of a terrorist, these agents can cause great psychological harm and social disruption as well.[2] Of the agents that can cause human disease, all except VEE and *C. burnetii* could be effectively transmitted through contaminated food and water.[3]

Classical methods for identifying biological agents that cause human disease have been used for more than 100 years and are well established.[4] These methods rely on agent cultivation, taking between 3 and 30 days, and require experienced personnel working in a well-equipped laboratory. In order to achieve agent identification with a high level of confidence, a combination of state-of-the-art immunological and nucleic acid analyses methods, as well as classical microbiological approaches, are needed to identify unknown biological agents.[5] Methods based on the immuno-magnetic-electrochemiluminescence assay for antigen detection and on polymerase chain reaction (PCR) gene amplification for nucleic acid detection are among the most sensitive.[6,7]

IMMUNOLOGICAL DETECTION PROTOCOLS

Electrochemiluminescence (ECL) is a process by which light is generated from a voltage-dependent, cyclic oxidation-reduction reaction of ruthenium heavy metal chelate. In the presence of tripropylamine (TPA), the redox reaction triggers the re-

---

[c]Address correspondence to: Dr. James A. Higgins, USDA-ARS, Rm. 202, Bldg. 1040, 10300 Baltimore Blvd., Beltsville, MD 20705; Telephone: 301-504-6443; Fax: 301-504-5306.
e-mail: jhiggins@lpsi.barc.usda.gov

TABLE 1. ECL assay sensitivities for botulinum toxin type A, C fragment in various biological matrices.

Matrix	Sensitivity (tested)	Sensitivity (absolute)
PBS/Tween-20	1 pg/ml	50 fg
Skim milk	10 fg/ml	0.5 fg
Serum	10 fg/ml	0.5 fg
Urine	100 fg/ml	5 fg

lease of photons that are detected and quantified within a photomultiplier tube. Ruthenium is used in a low molecular weight, heme-like form that can be easily conjugated to any protein using standard N-hydroxysuccinimide (NHS) ester binding chemistries. In the conjugated form, ruthenium serves as an ideal tracer molecule because it has little or no influence on antibody-antigen interactions. Detection is facilitated in a homogeneous assay format using 2.8 μm magnetic beads that serve as the solid substrate for the detection reaction. The beads also provide both the basis for the separation of antibody-antigen complexes from potential non-specific reactants and the means to bring the specific reactants into the proximity of the electrode. A potential of only 2 volts across the electrode is required to initiate the ruthenium redox reaction. As a result of this low potential, only those ruthenium atoms within 30–50 nm of the electrode are detected, reducing the effect of non-specific reactants present in the assay suspension.

An ECL detection system consists of an analyzer and a personal computer with menu-driven ORIOS software. The system's strengths come from its speed, sensitivity, accuracy, and precision over a wide dynamic range. In a typical agent-detection assay, magnetic beads conjugated to a capture antibody and a ruthenium-conjugated detector antibody are added to unknown sample. The analyzer draws the processed sample from a vortexing carousel, captures and washes the magnetic beads, and quantifies the electrochemiluminescent signal. The current system is automatable, uses stable reagents, and is highly sensitive into the 0.1–1 pg/ml range. The total assay time with the current system is approximately 30 minutes, plus up to one minute per sample reading time. The next-generation ECL analyzers will include both hand-held and high-throughput devices, making testing more accessible to caregivers and further decreasing the time required for testing and identification.

We have demonstrated the effectiveness of the ECL system for the detection of *Staphylococcus* enterotoxin B (SEB), ricin toxin, *Yersinia pestis* F1 antigen, *Bacillus anthracis* PA antigen, and Venezuelan equine encephalitis (VEE) virus. The technology could potentially be used with any biological agent and is limited only by the availability of high quality, high affinity antibodies or other ligands that can be used in the assay. This is of particular importance with regard to the detection of toxins per se, which are proteins and therefore not amenable to nucleic acid–based detection assays. In our hands, the technology is capable of detecting botulinum toxin at biologically significant concentrations (as low as 0.5 fg) for the first time. Representative results demonstrating the sensitivity of the botulinum toxin-detection assay are presented in FIGURE 1. As is the case with all agents tested, assay sensitivities vary between sample matrices (TABLE 1). Matrix-specific positive and negative control samples are used to establish standard curves and cutoff values. Using this ap-

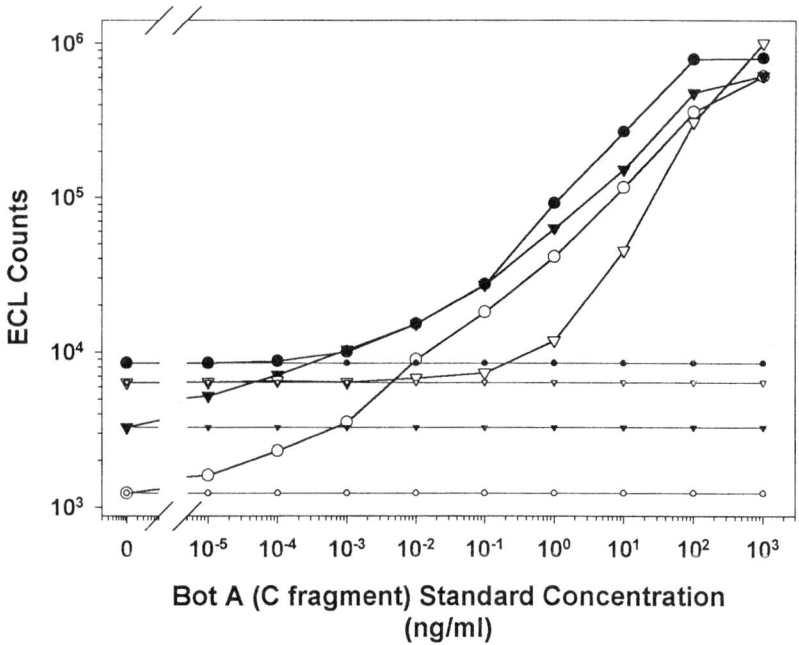

FIGURE 1. Standard curves generated from ECL analysis of various matrices spiked with Botulinum toxin type A, C fragment. *Closed circles* represent experiments completed in phosphate-buffered saline with 0.3% Tween-20; *open circles*, in skim milk; *closed triangles*, in serum; *open triangles*, in urine. Cutoff values are shown using narrow lines and small symbols, and were calculated from the mean of the results obtained from negative control samples, plus 2.5 standard deviations.

proach, any biological or environmental matrix can be tested and integrated into the testing platform. Using standard curves generated from test controls, the assays can be made fully quantifiable, increasing the utility of the technology even further. Such capabilities would be of great value when determining if a biological toxin is present in a food item or agricultural commodity at concentrations in excess of those expected from incidental contamination, i.e., a deliberate introduction.

The only major limitations of the ECL assay system revolve around the necessity for high quality reagents. Both detector and capture antibodies should ideally be highly specific and have high affinities for their respective antigens. All antibodies need to be highly purified to facilitate their efficient coupling to magnetic beads and ruthenium. However, once appropriate reagents have been identified and prepared, they are extremely stable, are easily lyophilized, and can be readily transported for field use. ECL-based agent identification is by far the most sensitive affinity-based diagnostic technology that we have tested to date. As the ECL hardware develops, this technology will prove itself to be an important component of an integrated approach to identification of potential biological weapons, both in the lab and in the field.

NUCLEIC ACID–BASED DETECTION PROTOCOLS

A variety of nucleic acid–based assays are currently available for detecting pathogenic microorganisms: the ligase chain reaction, the Qβ replicase-based system, the nucleic acid sequence–based amplification (NASBA) system, and PCR, among others.[8] Of these, PCR has been most extensively used for detecting DNA (deoxyribonucleic acid), and/or RNA (ribonucleic acid) unique to possible threat agents.

This technique involves the use of oligonucleotide primers (flanking a gene segment of interest), in conjunction with a heat-stable enzyme, e.g., *Taq* polymerase. By repeated cycling of the reaction through a range of temperatures permissible to the activity of the *Taq* polymerase, it is possible to synthesize large quantities of the desired gene segment to the exclusion of all other genes present in the sample.[7] The newly synthesized gene segments, or DNA molecules, are referred to as amplicons and can be analyzed and viewed by electrophoresis on agarose gels. When stained with the chemical dye ethidium bromide, amplicons are revealed as bands of fluorescing material in the gel. Because the investigator determines beforehand both the identity and length of the gene segment he or she is interested in amplifying, the size of the observed bands in the gel is indicative of an accurate and successful assay. It is possible to simultaneously assay a sample by PCR using as targets various gene segments, unique to a species or strain of microorganism. These can be used to both improve the detection limit and to determine if the organism in question contains particular genes associated with virulence or pathogenicity.

Because of its ability to amplify DNA from extremely small quantities of starting material, PCR offers DNA detection limits in the femtogram or even attogram range. Because it allows the investigator to design primers unique to gene segments of interest, PCR also offers a high level of specificity, with discrimination possible at the level of individual nucleotides, particularly when oligonucleotide probes (see below) are included in the assay. The PCR assay is also relatively quick to perform and analyze compared to culture and isolation techniques—depending on the need for sample preparation, results can be obtained in several hours.

PCR does have disadvantages. The assay requires pure and undegraded nucleic acids (templates) for optimum performance, making the removal of inhibitory substances, which are present in many clinical and environmental samples, a necessity.[9] Additionally, because of its exquisite sensitivity, contamination of PCR reactions with previously generated amplicons can result in false-positive assays. Avoiding contamination problems involves the use of specially designated laboratory equipment and handling procedures. Finally, successful use of PCR requires more involved training of personnel and the provision of properly equipped laboratories, with attendant requirements for larger budgets.

Sample Preparation

The minimum requirements for a sample preparation procedure for preparing PCR-amplifiable nucleic acid are (1) the release of the nucleic acid from the target organism in sufficient quantity and quality to support the *in vitro* amplification process and (2) the removal of PCR inhibitory contaminants arising from the sample matrix.[9] These requirements are influenced by a number of factors. Cell type is an important consideration. For example, it is more difficult to prepare DNA from *Ba-*

cillus anthracis spores than from vegetative cells under any circumstances. Another factor is the amount and type of PCR inhibitory contaminants present in the sample matrix. For example, it is more difficult to prepare DNA from *B. anthracis* vegetative cells from blood or soil samples than buffer samples. Furthermore, it is more difficult to prepare DNA from organic soils than from sandy soils. Finally, the type of nucleic acid being purified influences the sample preparation strategy. The ubiquity of ribonucleases (RNases) and their resistance to inactivation has to be taken into consideration when pathogens with RNA genomes or when RNA transcripts from an organism with a DNA genome are being targeted. Other features of a sample preparation procedure that are desirable and make it more likely that PCR will be routinely used in diagnostic laboratories are minimal use of hazardous materials, such as phenol, and maximum opportunity for automation to minimize the labor-intensive steps that hinder current sample preparation procedures.

There are a number of commercially available procedures for preparing nucleic acid for PCR. A successful approach for preparing DNA for PCR is based on variations of the Boom procedure.[10] In this method, organisms and their nucleic acid–protein complexes are lysed and disassociated with a chaotropic agent. Nucleic acid is then preferentially adsorbed to silica under high salt conditions, contaminants removed by selective washes, and DNA preferentially eluted with a low salt buffer. Qiagen, Inc. (Chatsworth, CA) produces a variation of this procedure in two forms: the QIAamp® Tissue Kit and the QIAamp® Blood Kit. The Qiagen procedures incorporate a protease digestion step to enhance the chaotropic salt–mediated lysis of cells. These two procedures are identical except for the protease digestion step. In the tissue kit procedure, there is a 60-min incubation at 55°C with proteinase K, before a 10-minute incubation at 70°C in the chaotropic salt–containing lysis buffer. The blood kit procedure combines the protease digestion step with the chaotropic disruption step during the 10-min incubation at 70°C. Instead of proteinase K, the procedure uses Qiagen protease, a Qiagen product optimized specifically for preparing DNA from blood. We found that the tissue kit procedure was better for preparing spore samples in buffer, i.e., the endpoint detection limit for it was greater than for the blood kit, 4×10^3 colony-forming units (cfu)/40 μl sample versus 4×10^4 cfu/40 μl sample. We speculated that the proteinase K digestion step was more efficient for preparing spores for DNA extraction by this method. However, we also found that the most sensitive method for preparing spores for DNA extraction was to germinate them for 60 minutes. After this treatment the spores behaved like vegetative cells and we were able to detect $4 \times 10^1 - 4 \times 10^2$ cfu/40 μl sample.

Although the procedure is labor-intensive, with multiple centrifugation steps, and is therefore not readily adaptable to automation, it is very reliable and sensitive. It is also independent of the sample matrix. We found we could prepare *B. anthracis* vegetative cells with equal endpoint detection limits, from whole blood, plasma, or serum (FIG. 2). Preliminary experiments with *Yersinia pestis*, the etiologic agent of plague, and *Brucella abortis*, the etiologic agent of brucellosis, indicate that the Qiagen procedures will also be effective in preparing these agents for PCR.

In an effort to take advantage of the efficacy of this technology and to minimize its labor-intensive shortcomings, we are evaluating the Autolyser®, an instrument developed by XOHOX Research Institute (Menlo Park, CA). It fully automates this process. The Autolyser can be transported in a suitcase, complete with reagents and

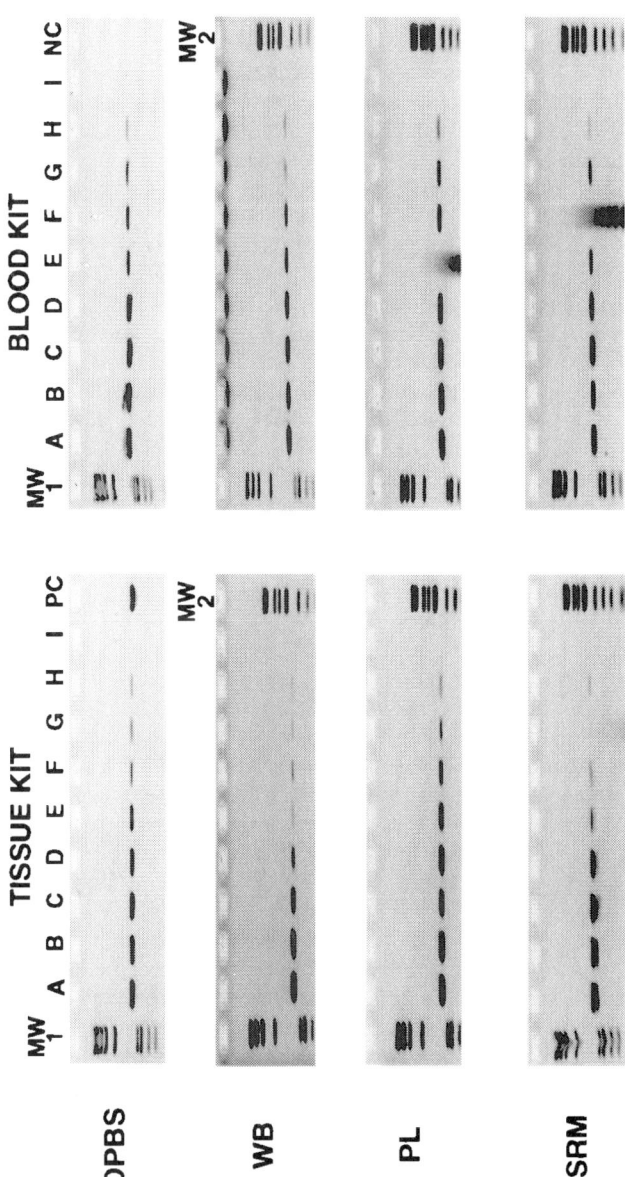

FIGURE 2. Comparison of tissue kit and blood kit procedures for preparing samples diluted in Dulbecco's phosphate-buffered saline, whole blood, plasma, and serum. Serial dilutions of γ-irradiated *Bacillus anthracis* vegetative cells were prepared in Dulbecco's phosphate-buffered saline (DPBS), whole blood (WB), plasma (PL), and serum (SRM) and processed by either the tissue kit or blood kit procedure. Colony-forming units/40 µl sample: (A) 5.0×10^4; (B) 2.5×10^4; (C) 1.2×10^4; (D) 6.2×10^3; (E) 3.1×10^3; (F) 1.5×10^3; (G) 7.8×10^2; (H) 3.9×10^2. Sample I is a no sample, diluent control. PC = PCR positive control; NC = PCR negative, no template control. MW1 = ΦX174 *Hae*III molecular weight markers; MW 2 = ΦX174 *Hin*fI molecular weight markers.

a laptop computer with menu-driven software. After a sample is introduced into a disposable extractor device, the instrument automatically dispenses lysis, wash, and elution reagents according to user-modifiable software instructions. Purified nucleic acid is deposited into a sterile vial approximately 15 minutes after sample introduction. The procedure requires no additional laboratory equipment and requires minimal training to operate. With this instrument we have successfully prepared PCR-amplifiable DNA from *B. anthracis* spores and vegetative cells.[11] Currently, the endpoint detection limit of PCR performed on Autolyser-extracted DNA is not as great as for DNA extracted with Qiagen systems; however, the Autolyser's ease of use justifies further development.

We have also evaluated two relatively new technologies based on modifications of Schleicher and Schuell 903 paper: FTA® paper, manufactured by Flinders Technologies, Inc. (Adelaide, South Australia), and distributed by Fitzco, Co. (Maple Plain, MN) and Life Technologies, Inc. (Bethesda, MD); and IsoCode® Cards, manufactured and distributed by Schleicher and Schuell (Keene, NH). Both procedures are optimized for preparing DNA from whole blood and employ proprietary chemistries that remove or inactivate blood-specific PCR inhibitors. In the FTA procedure, a sample is spotted onto the paper and dried. A 3-mm punch from the paper is processed through a series of washes with a specific FTA purification reagent, Tris EDTA buffer with 0.1 mM EDTA, and ethanol, to remove inhibitors. The processed punch with the captured DNA is directly used in the PCR reaction. With the IsoCode procedure, the sample is also spotted, dried, and punched, but is subjected to a single H_2O wash. After the wash the DNA is eluted from the punch by heating to 95–100°C for 30–45 minutes. The eluate and paper both contain amplifiable DNA. Experiments are in progress to determine whether we can eliminate the elution step without adversely affecting sensitivity. Preliminary experiments indicated that both the FTA and IsoCode procedures were approximately equally sensitive for preparing DNA from *B. anthracis* vegetative cell from buffer and whole blood samples. However, in our hands, FTA paper was significantly less efficient in preparing *B. anthracis* DNA from either plasma or serum. Because of this deficiency and because of the relative simplicity of the IsoCode procedure—one water wash, versus multiple washes with different reagents—we decided to further evaluate the IsoCode paper procedure as a means for preparing DNA for PCR. We found the procedure to be very effective, enabling us to detect 2×10^2 cfu/20 μl sample of germinated *B. anthracis* spores or vegetative cells. Like the Qiagen procedures, detection was independent of sample matrix (FIG. 3).

For RNA targets (i.e., viruses), one of the most effective methods for preparing RNA for amplification by the reverse transcriptase (RT)-PCR procedure is a variation of the Chomczynski and Sacchi procedure[12] in which cells are lysed and protein–nucleic acid complexes dissociated by a chaotropic agent, such as guanidinium isothiocyanate (GITC). The RNA is then extracted with acidic phenol–chloroform–isoamyl alcohol and concentrated by alcohol precipitation. In addition to being an effective agent for lysing cells and disassociating the RNA–protein complexes, GITC effectively inactivates RNases, satisfying an essential requirement for preparing RNA targets. There are a number of commercially available systems that employ procedures based on this technology. We have tested the RNAgents™ System from Promega, Inc., (Madison, WI) and a variation of the TRIzol procedure (Life Tech-

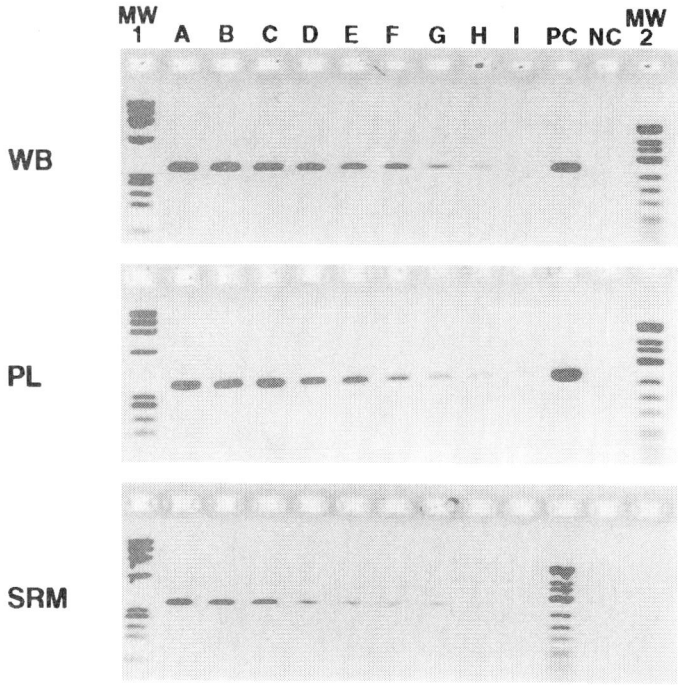

FIGURE 3. Comparison of IsoCode® paper for preparing vegetative cells diluted in whole blood, plasma, or serum. Gamma-irradiated *Bacillus anthracis* vegetative cells were diluted in whole blood (WB), plasma (PL), and serum (SRM) and prepared by the IsoCode procedure. Colony-forming units/20 µl sample: (A) 2.0×10^5; (B) 1.0×10^5; (C) 5.0×10^4; (D) 2.5×10^4; (E) 1.2×10^4; (F) 6.2×10^3; (G) 3.1×10^3; and (H) 1.5×10^3. I is a no sample, diluent control. PC = positive PCR template control; NC = negative PCR no template control. MW1 = ΦX174 *Hae*III molecular weight markers; MW2 = ΦX174 *Hin*fI molecular weight markers.

nologies, Inc., Bethesda, MD) in which the GITC and phenol are combined into a single reagent instead of two separate ones.[13] This procedure is very effective for preparing RNA from RNA-genome viruses, such as Rift Valley fever virus (RVF), or Venezuelan equine encephalitis virus (VEE), enabling us to detect between 200–1,500 plaque-forming units (pfu)/100 µl, depending on experimental conditions. However, the procedure is labor-intensive, not readily adaptable to automation, and uses hazardous chemicals that require special handling and disposal.

Another procedure we have investigated for isolating nucleic acid is immunomagnetic-separation (IMS). In this procedure, paramagnetic beads are derivitized with a specific antibody. These derivitized beads are used to capture a select organism from the sample milieu in much the same way as capture antibodies are used to immobilize specific antigenic targets onto microtiter plates for ELISA procedures. The advantages of this procedure are that it can effectively remove the target organism from the sample milieu, thus eliminating potential PCR inhibitory contaminants

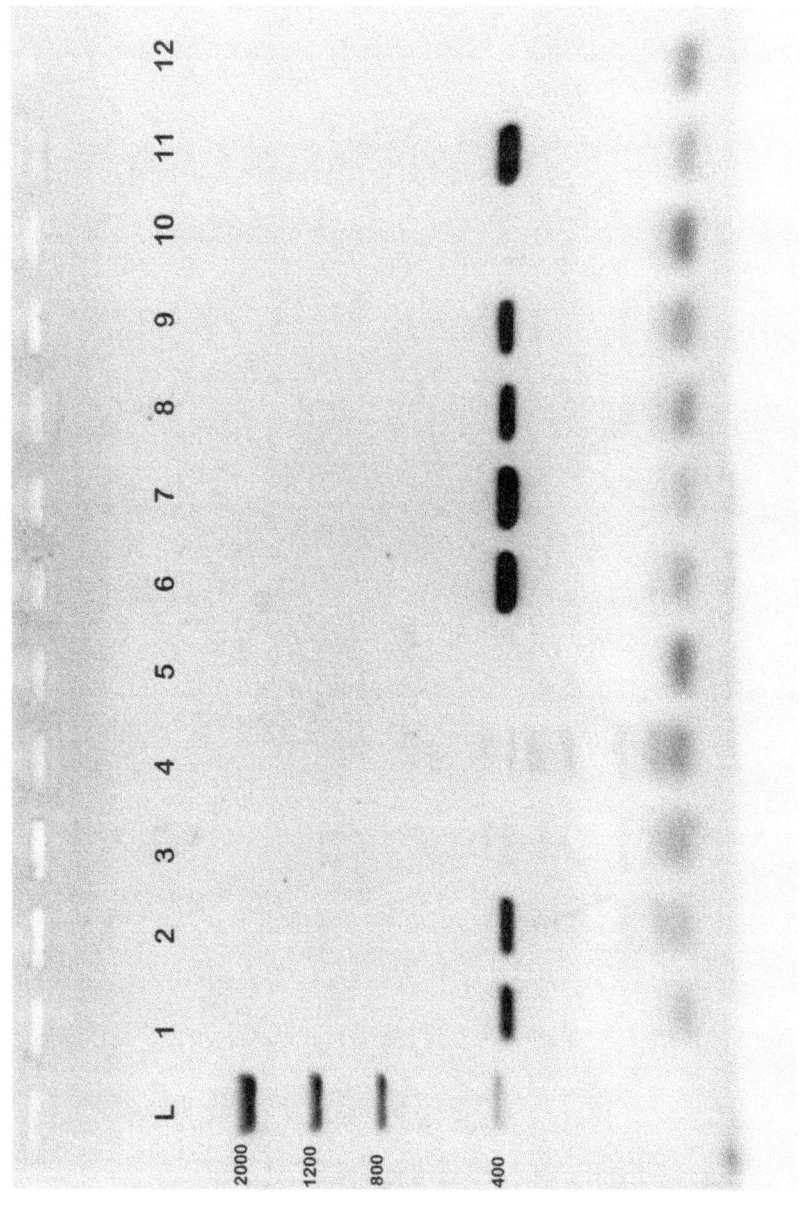

in the process. At the same time, because of the high surface area to volume ratio of the beads, this procedure can also concentrate the analyte. Moreover, this procedure has the potential for automation. We have developed IMS procedures for VEE and have found that this technique increased the sensitivity of subsequent RT-PCR, enabling us to detect 25–100 pfu/100 µl sample, equivalent to an eightfold increase in detection limit over the GITC–acidic phenol extraction procedure. We have also developed an IMS procedure for two bacteria, *Yersinia pestis* and *Brucella abortis*, and found that DNA isolated with this method yielded PCR detection limits at least as sensitive as DNA isolated with the Qiagen procedures. The IMS extraction procedure had the added advantages of being able to concentrate dilute targets from larger volumes and more amenable to automated operation.

As an example of how rapid sample preparation methods and PCR might be applied to the investigation of a food item or an agricultural material suspected of harboring a biowarfare agent, we performed an experiment in which either a 1-ml or 10-µl aliquot of whole milk was spiked with serial log dilutions of *Bacillus anthracis* Ames vegetative cells. We used these two volumes for our spiking experiment to determine if the inhibitory action of lipids or proteins present in the milk could be removed using the Isocode sample preparation method. After addition of the *B. anthracis*, the milk was vortexed to disperse the bacteria thoroughly, and a 10-µl portion was spotted onto Isocode disks (about 6 mm in diameter) and processed. A 10-µl sample of the eluate was used as template for *B. anthracis* protective antigen gene amplification.

Results are shown in FIGURE 4. When 10-µl aliquots from the 1-ml spiked milk samples were assayed (lanes 1–5), the detection limit was approximately 1,000 cfu per aliquot (Lane 2). When smaller portions (10 µl) of milk were spiked with *B. anthracis*, the detection limit of the assay was approximately 100 cfu per aliquot (Lane 9). This assay indicates that it is possible to detect *B. anthracis* in milk at concentrations that may be encountered in a possible bioterrorism scenario using an inexpensive, simple sample preparation technique. It should be noted that, as with other PCR assays, the detection limit can undoubtedly be improved by the use of either a nested PCR assay or a PCR-enzyme immunoassay (see below); but these techniques can add several hours to the overall assay time.

PCR-ENZYME IMMUNOASSAY

The PCR-enzyme immunoassay (PCR-EIA) is a relatively recent development in PCR technology. The basic assay reagents are commercially available (Boerhinger Mannheim). The assay utilizes the incorporation of an altered nucleotide, digoxige-

---

FIGURE 4. Detection of *Bacillus anthracis* Ames vegetative cells in whole milk. One-ml and 10-µl volumes of milk were spiked with a dilution series of colony-forming units (cfu), and 10-µl aliquots spotted onto Isocode® paper and processed for PCR with the BAN 23 primer set. Lane L: DNA mass ladder with sizes of the rungs indicated. Lanes 1–4: aliquots from 1 ml milk spiked with 2×10^5, 1×10^5, 1×10^4, and 1×10^3 cfu, respectively. Lane 5: unspiked milk control. Lanes 6–9: aliquots from 10 µl milk spiked with 2×10^4, 1×10^4, 1×10^3, and 1×10^2 cfu, respectively. Lane 10: unspiked milk control. Lane 11: *B. anthracis*–positive control. Lane 12: no template control.

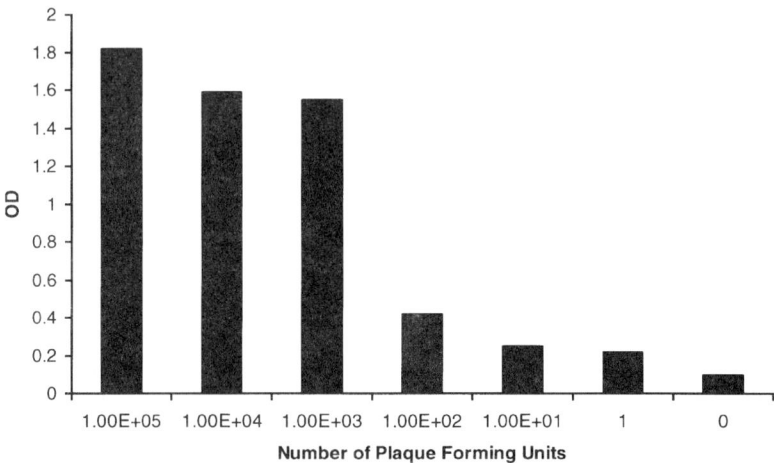

FIGURE 5. Detection limit of PCR-EIA for Venezuelan equine encephalitis virus. The optical density units are plotted on the y axis, and the number of plaque-forming units of VEE are plotted on the x axis.

nin-11-deoxyuridine triphosphate (d11-dUTP), into amplicons. These are in turn subjected to hybridization with a biotin-labeled oligonucleotide probe whose sequence is determined by the investigator. Probe bound to target regions on the amplicons is visualized by a colorimetric enzymatic reaction, mediated by an antibody and a colorimetric substrate. The assay is performed using a microtiter plate format, with positive reactions indicated by color change, in proportion to the quantity of template in the samples. A successful assay can detect femtogram quantities of pathogen DNA. The assay allows many samples (up to 96, including controls) to be processed at once. The use of a biotin-labeled, oligonucleotide probe improves specificity compared to conventional PCR. The drawbacks of PCR-EIA are that it requires between 2–6 h to complete and involves multiple pipetting steps. We have developed PCR-EIA assays for the detection of bacterial and viral threat agents. An example is depicted in FIGURE 5; here, the PCR-EIA allows detection of as little as 1 pfu of VEE virus.

Fluorogenic Probe-Based PCR

These techniques incorporate fluorescent dye molecules either as stains for the amplicons themselves or attached to oligonucleotide probes, which then hybridize to the amplicons, into the PCR reaction. The advantage of using fluorogenic markers in PCR is that with the appropriate detection platforms, the accumulation of amplicons can be monitored in "real time," i.e., as the reaction(s) progress. Data can be presented as amplification plots on a graph, obviating the need for gel electrophoresis to determine if a given sample is positive. A number of different assay systems and platforms are commercially available. One such system is TaqMan® 5′ nuclease assay (5NA).[14]

The TaqMan 5NA (Perkin Elmer/Applied Biosystems) utilizes an oligonucleotide probe, whose sequence is selected by the investigator, double-labeled with fluorescent reporter and quencher dye molecules. As amplicons accumulate during the course of the reaction, the probe will hybridize to any target sequence present on the amplicons. When exposed to pulses of intense light (e.g., a laser), the fluorescent dyes located on the probe will respond with characteristic emission spectra. These spectra can be monitored by an appropriate platform and data collated and presented to the user as amplification plots.[15] Because different probes can be labeled with different dye molecules, it is possible to simultaneously assay the same sample for the presence of different target sequences (i.e., "multiplexing"). In practical terms, this would allow the same sample to be assayed for the presence of several different threat agents in a one-tube reaction.

The main advantage offered by the 5NA is that with real-time monitoring, results can be obtained much quicker than with conventional PCR or PCR-EIA. It is also possible to design probes that allow discrimination between single nucleotide differences in gene sequences. As demonstrated below, this can be of importance in differentiating between pathogenic and nonpathogenic microorganisms.

The use of probe-based, nucleic acid amplification protocols also provides investigators with the ability to identify "artificially created" pathogens, which, unfortunately, are feasible using modern molecular biology techniques. An experienced molecular biologist could insert genes coding for toxins, virulence factors, or antibiotic-resistance proteins into bacterial and viral threat agents.

To address the possibility of encountering pathogens that have been "fortified" with virulence genes from other unallied microorganisms, USAMRIID's diagnostic research is focusing on rapid detection and identification of a number of known virulence genes for a variety of pathogens. It is feasible, for example, to perform molecular tests on a given bacterial or viral isolate for the presence of introduced toxin genes, which would not normally be detected in conventional assays. As research into these detection capabilities proceeds, it is hoped that it will provide authorities with the ability to rapidly determine if a given isolate is deliberately altered at the genetic level to enhance its use as a biological weapon.

We have functional TaqMan 5NA for a variety of bacterial and viral threat agents: *B. anthracis*, *Y. pestis*, *Brucella* spp., *F. tularensis*, *Staphylococcus* enterotoxin B gene, Venezuelan equine encephalitis virus, and a number of orthopoxvirus species. The *Y. pestis* assay has been successfully used to detect bacterial DNA in oropharyngeal swabs taken from monkeys infected with aerosolized formulations of *Y. pestis* and in artificially infected vector fleas.[16] Our orthopoxvirus assays allow differentiation of strains at the single-nucleotide level.[17,18] Because of the high degree of sequence homology among the orthopoxviruses, this capability is vital to determine if a virus isolate is pathogenic for humans or is an animal virus whose presence in the environment is otherwise unremarkable.

An example of the use of rapid, probe-based PCR methods to identify material from an actual incident involving a potential threat agent is provided in FIGURE 6. A law enforcement agency provided the Special Pathogens Department at the Diagnostic Systems Division with a bacterial preparation suspected of containing *B. anthracis*. Bacterial colonies cultured from the confiscated material were assayed by 5NA, using probes to the protective and capsular antigens of *B. anthracis*. In FIGURE 6, the

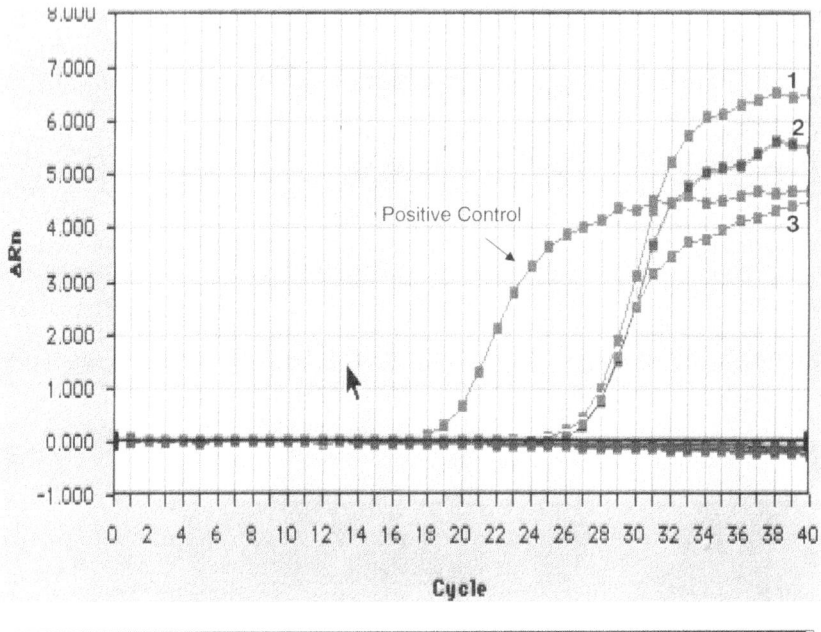
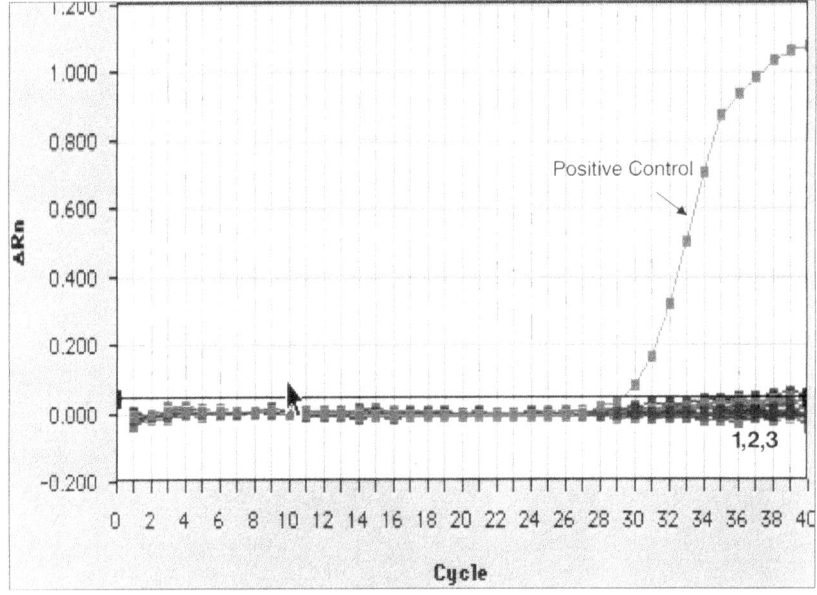

FIGURE 6. Fluorogenic probe-based PCR for *Bacillus anthracis*. The ΔRn values are plotted on the y axis, and the PCR cycle numbers are plotted on the x axis. The plot for the positive control sample is indicated; the plots for the unknown samples ($N = 3$) are designated by number. Amplification plots located below the threshold line (ΔRn = 0) are considered negative, and include no template controls. (A, top) The unknown samples were assayed with primers and probe for the *B. anthracis* protective antigen gene. (B, bottom) The same samples were assayed with primers and probe for the *B. anthracis* capsular antigen gene.

amplification curves for each sample assayed are plotted on a graph. The ΔRn values (y axis) are indicative of probe activity; samples with plots below the threshold line (ΔRn = 0) are considered negative. The samples were positive when assayed with primers and probe for the protective antigen (FIG. 6A). However, when assayed with primers and probe for the capsular antigen gene, a loci associated with virulent strains of anthrax, the samples were negative (FIG. 6B). The conclusion from this and other data was that the sample was an avirulent strain of *B. anthracis* conventionally used in vaccine preparations. The assay demonstrated that the sample lacked the capsular antigen gene, which is associated with virulent strains of *B. anthracis*. Results such as these, which were obtained within 2 h of beginning the 5NA, can have obvious implications in determining safety measures with regard to exposure (accidental or otherwise) to such a bacterial preparation. Information from 5NA analyses of confiscated material can also be a useful component of forensic investigations.

FUTURE DIRECTIONS FOR NUCLEIC ACID–BASED DETECTION ASSAYS

A significant disadvantage of fluorogenic probe assays is the requirement for expensive, bulky platforms for real-time monitoring of the reactions. This lack of portability can hamper the ability of real-time PCR to be carried out in field situations. Accordingly, the Department of Defense, in conjunction with investigators at the Department of Energy's Lawrence Livermore National Laboratories, has sponsored research into the development of miniature analytical thermal cyclers. One such platform is the MATCI (miniature analytical thermal cycler instrument) manufactured by Lawrence Livermore National Laboratory.[18,19] The MATCI utilizes silicon wafers to mediate the heating and cooling of the reaction tube, in conjunction with light-emitting diode (LED) stimulation of fluorogenic probes present in the PCR tube.[19] The entire apparatus is suitcase-size, portable, and offers "true" real-time monitoring of probe activity (i.e., the amplification plots onscreen are continually updated cycle by cycle). The MATCI has been successfully used to detect, and differentiate between, orthopox virus species.[18]

An improved version of this platform is designated the Automatic Nucleic Acid Analyzer (ANAA) and features ten independently programmable heating blocks.[20,21] When *Bacillus subtilis* spores (used to simulate *B. anthracis* spores) were assayed by fluorogenic probe–based PCR in this instrument, positive signals were viewed in 18–26 minutes.[20] A handheld platform offering equivalent real-time monitoring capabilities, with four blocks, is in the prototype stage and may be affordable to various investigators and laboratories by the year 2000 (P. Belgrader, Lawrence Livermore National Laboratories, personal communication).

Other Department of Energy–sponsored biological weapon detection technologies include the MiniFlo cytometer, which uses immunofluorescence-based sensors to detect both pathogens and toxins. In field exercises held in 1996 at Dugway, Utah, 1,600 analyses on 400 samples yielded an 87% positivity rate, with less than a 0.5% false positive rate.[21]

Researchers at Cepheid, a private firm in Sunnyvale, CA, are also constructing portable thermal cyclers with real-time capabilities with the goal of marketing hand-held platforms offering combined microfluidics-based sample preparation, thermal cycling, and data presentation functions. The reaction chamber heating and optics functions would be mediated via the use of a specially designed ceramic chip ("I-CORE"). The Smartcycler™, offering real-time, fluorogenic probe-based PCR, is expected to be marketed within the year. Ultimately, a Cepheid system would be capable of accommodating sample volumes as large as 1–3 ml. This is of importance when few pathogens are expected to be present in the sample. The company plans to develop a briefcase-size thermal cycler unit with integrated sample processing; the instrument would be available within the next several years and priced to be affordable to a variety of end users.[22,23]

We have demonstrated that our orthopoxvirus 5NA is readily adaptable for use on these novel platforms, performing as well as or better than commercial instruments.[18] Ongoing research in our laboratory has indicated that with such platforms a nucleic acid–based detection procedure—from sample preparation, amplification, and determination of results—can take less than one hour.[18]

APPLYING ADVANCED DIAGNOSTIC TECHNIQUES IN BIOTERRORISM SCENARIOS

How feasible is it to move these sophisticated laboratory instruments and assays from the bench to the field, from an essentially reactive to a proactive role? The U.S. Army has been deploying rapid diagnostics capabilities in the field since the 1980s. More recently, during Operation Vigilant Warrior (1994), we demonstrated for the first time that advanced molecular diagnostics using PCR can be performed under field conditions for identification of disease agents. Subsequently, the 520th Theater Area Medical Laboratory (TAML, currently commanded by Colonel William Chambers, and stationed at Aberdeen Proving Ground in Maryland) was activated. TAML is a self-contained laboratory that can be airlifted to overseas locations. It provides theater-level laboratory support for preventive medicine activities, using a variety of sophisticated diagnostic assays. The laboratory is staffed by enlisted personnel and has demonstrated an ability to conduct diagnostic analyses with a high degree of accuracy and reproducibility over several months under field conditions. For example, the 520th TAML was recently deployed to an overseas location in support of U.S. Army personnel. Despite difficult environmental conditions (high ambient temperatures that made it necessary to perform the bulk of the assays after sunset, sand infiltration of equipment, etc.) the laboratory conducted numerous detection assays each day. Results obtained from the TAML assays were confirmed at USAMRIID, indicating that TAML operations represent a highly successful transfer of technology from the research laboratory to the field (T. Cao and F. Knauert, unpublished data). Therefore, it is feasible to envision a civilian counterpart with the same capabilities as the 520th TAML.

How would rapid detection techniques such as the ones described above be used in the event of intentional contamination of food or water with pathogenic agents? Because such incidents are rare, it is necessary to extrapolate from what we do know occurred in the United States in The Dalles, Oregon, in 1984. There, a large outbreak of salmonellosis among patrons and workers at 10 restaurants was ultimately ascribed to deliberate contamination of salad bar foods by members of the Bhagwan Shree Rajneesh religious cult. In addition to the restaurants, cult members also contaminated produce at a grocery store and intended to contaminate the municipal water supply as well. Evidently the cult had hoped to cause enough widespread illness to significantly lower voter turnout on a land-use balloting issue that would have hampered the cult's activities. It took more than a year for investigators to accumulate enough evidence to identify the cult as the source of the outbreak. This was aided by seizure of a *Salmonella* culture from the Rajneesh Medical Center. A panel of biochemical and genetic assays indicated that it was identical to the strain isolated from outbreak patients.[24] At the time of the outbreak and investigation (1984–1985), PCR-based techniques were not yet invented

The Oregon outbreak ultimately involved at least 751 cases, with 45 people ill enough to be admitted to the hospital. It is troubling to realize that while the cult had selected a relatively benign pathogen, it was easily obtained from a commercial microbiological specimen provider; it was cultured in the Rajneesh laboratory without the need for special facilities and equipment; and it was spread simply and easily throughout the community without attracting any undue suspicion. Soon after the epidemiologic investigations began, the authorities did have suspicions that the outbreak was intentional, but for various reasons, this hypothesis was not immediately pursued.[24]

More recently, in 1996, in a Texas medical center laboratory, 12 workers were diagnosed with *Shigella dysenteriae* infection. The source of the infection was traced to pastries anonymously left in a common break room. Because the facility routinely performed microbiological assays, bacterial stocks and culturing equipment were kept on the premises. Investigators examined stool sample isolates and bacterial stocks from the lab storage freezer. Using pulse-field gel electrophoresis, a technique that allows comparative analysis of chromosomal DNA profiles of different bacterial strains, they determined that the *S. dysenteriae* strains from the freezer and stool samples were identical. The investigators concluded that the pastries had been deliberately contaminated and a criminal investigation was begun.[25]

In both incidents, the realization that deliberate contamination of food items was the source of infection happened only after the outbreaks had occurred. This underscores how difficult it can be to prevent or ameliorate illnesses caused by exposure to covertly contaminated food or water sources.[26] The use of gastrointestinal pathogens, which are relatively easy to acquire, can confound attempts to determine if the outbreak was triggered by natural or intentional activities. For example, it would require an intensive effort to distinguish between an outbreak caused by naturally present botulinum toxin and one caused by a deliberately introduced formulation.

Alternatively, since biowarfare agents such as anthrax or encephalitis viruses would constitute more exotic pathogens, for many health departments confronted with unusual symptoms among a population exposed to contaminated food or water, arriving at a timely and accurate diagnosis would be difficult unless select medical personnel had previous experience with illness caused by these agents. The symptoms of gastrointestinal anthrax, an extremely rare disease in nature, are similar to those of most gastrointestinal tract infections, with fever, diarrhea, nausea, and vomiting predominating before the infection progresses to acute abdominal pain, at which time the infection is most serious and mortality rates can approach 50%.[27]

The utility of rapid detection assays would therefore seem to address two areas: (1) prophylactic monitoring of food or water suspected of being the target of a bioterrorist attack and (2) serving as "first use" diagnostics when an otherwise routine outbreak of gastrointestinal illness shows evidence of being something else entirely. (We wish to stress that rapid detection techniques would certainly not be used alone in any effort, but would be used in conjunction with more traditional techniques, such as culturing of suspected agents, animal assays, as well as highly specific immunodiagnostic assays, such as ECL).

With regard to monitoring of food or water sources suspected of having been contaminated with biological warfare agents, we feel that the rapid sample preparation techniques and real-time diagnostic assays developed at USAMRIID would allow authorities to perform the quickest and most accurate tests to determine if the threat is real. We envision the on-site use of multiplex, probe-based PCR to simultaneously assay samples for the presence of a number of possible infectious agents, in conjunction with field-based immunoassays for confirmation of PCR procedures and detection of various toxins. Results of these assays would allow authorities to take appropriate countermeasures, such as securing the food or water source to prohibit its use, and also constitute evidence for criminal prosecution.

In the event of an outbreak of gastrointestinal illness suspected to have been caused by the deliberate introduction of biological warfare agents into food or water items, the use of real-time, rapid diagnostics would be of importance in establishing early diagnoses for optimum treatment and protective measures. Because most public health laboratories lack the reagents and expertise required to diagnose disease caused by agents such as poxviruses and anthrax or toxins such as ricin, timely access to a "package" of rapid detection instruments and assays would be critical.

In addition to the advantages described above, it may be helpful to think of rapid detection capabilities as contributing to a well-formulated and effective policy of deterrence of biological warfare/terrorism.[28] In conjunction with other countermeasures, such as drug prophylaxis, immunizations, protective clothing, and shelters, rapid diagnostics can help to minimize casualties and serve as a signal to potential adversaries that the United States is becoming less vulnerable to these attacks.

In conclusion, we are confident that rapid detection assays can provide crucial information to personnel confronted with an incident of suspected bioterrorism. Alternatively, these assays can be utilized in a proactive mode to monitor vulnerable commodities for safety and hygiene. Such assays could be conducted within hours of receiving samples, providing results more rapidly than techniques currently available in most diagnostic laboratories. The sensitivity and accuracy of rapid diagnostic techniques, augmented by traditional assays, would provide the abovementioned personnel with a valuable resource for use against biological warfare or terrorism aimed at the nation's food and water resources.

ACKNOWLEDGMENTS

The authors wish to thank Gerry Howe, Phil Craw, Michelle Shipley, John Spencer, Leo Azucenalagos, Tuan Cao, Terry Abshire, Jeff Teska, Cindy Rossi, Joe Mangiafico, Phil Belgrader, Ted Hadfield, and Anjeli Sonstegard for their assistance. The authors would also like to acknowledge the contributions to assay development made by past members of the Diagnostic Systems Division: George Korch, Roger Parker, Rich Lofts, and John Lowe.

REFERENCES

1. CHRISTOPHER, G.W., T.J. CIESLAK, J.W. PAVLIN & E.M. EITZEN, JR. 1997. Biological warfare: A historical perspective. J. Amer. Med. Assoc. **278:** 412–417.
2. HOLLOWAY, H.C., A.E. NORWOOD, C.S. FULLERTON, C.C. ENGEL, JR. & R.J. URSANO. 1997. The threat of biological weapons: prophylaxis and mitigation of psychological and social consequences. J. Amer. Med. Assoc. **278:** 425–438.
3. BENENSON, A., ED. 1995. Control of Communicable Diseases Manual. 16th edit. American Public Health Association. Washington, D.C.
4. MURRAY, P.R, E.J. BARON, M.A. PFALLER, F.C. TENOVER & R.H. YOLKEN, EDS. 1995. Manual of Clinical Microbiology. 6th edit. ASM Press. Washington, D.C.
5. FRANZ, D.R., P.B. JAHRLING, A.M. FRIEDLANDER, D.J. MCCLAIN, D.L. HOOVER, W.R. BRYNE, J.A. PAVLIN, G.W. CHRISTOPHER & E.M. EITZEN, JR. 1997. Clinical recognition and management of patients exposed to biological warfare agents. J. Amer. Med. Assoc. **278:** 399–411.
6. GATTO-MENKING, D.L., H. YU, J.G. BRUNO, M.T. GOODE, M. MILLER & A.W. ZULICH. 1995. Sensitive detection of biotoxoids and bacterial spores using an immunomagnetic-electrochemiluminescence sensor. Biosens. Bioelectron. **10:** 501–507.
7. MULLIS, K., F. FALOONA, S. SCHARF, R. SAIKI, G. HORN & H. ERLICH. 1986. Specific enzymatic amplification of DNA *in vitro*: The polymerase chain reaction. Cold Spring Harbor Symp. Quant. Biol. **51:** 263.
8. PERSING, D.H. 1993. *In vitro* nucleic acid amplification techniques. *In* Diagnostic Molecular Microbiology: Principles and Applications. D.H. Persing, T.F Smith, F.C. Tenover, & T.J. White, Eds.: 51–87. American Society for Microbiology. Washington, D.C.
9. WILSON, I.G. 1997. Inhibition and facilitation of nucleic acid amplification. Appl. Environ. Microbiol. **63:** 3741–3751.
10. BOOM, R., C.J. SOL, M.M. SALIMANS, C.L. JANSEN, P.M. WERTHEIM-VAN DILLEN & J. VAN DER NOORDAA. 1990. Rapid and simple method for purification of nucleic acids. J. Clin. Microbiol. **28:** 495–503.
11. FIELDS, R., F.K. KNAUERT, B.C. COURTNEY & E.A. HENCHAL. 1998. Fully automated purification of bacterial DNA in a sealed device with no pipetting or centrifugation steps. Presented at the Cambridge Healthtech Institute's DNA/RNA Diagnostics Conference, Washington, D.C. May 19–21, 1998.
12. CHOMCZYNSKI, P. & N. SACCHI. 1987. Single-step method of RNA isolation by acid guanidinium thiocyanate-phenol-chloroform extraction. Anal. Biochem. **162:** 156–159.
13. CHOMCZYNSKI, P. 1993. A reagent for the single-step simultaneous isolation of RNA, DNA, and proteins from cells and tissue samples. Bio/Techniques **15:** 532–534.
14. HOLLAND, P.M., R.D. ABRAMSON, R. WATSON & D.H. GELFAND. 1991. Detection of specific polymerase chain reaction product by utilizing the 5'-3' exonuclease activity of *Thermus aquaticus* DNA polymerase. Proc. Natl. Acad. Sci. USA **88:** 7276–7280.
15. HEID, C.A., J. STEVENS, K.J. LIVAK & P.M. WILLIAMS. 1996. Real time quantitative PCR. Genome Res. **6:** 995–1001.
16. HIGGINS, J.A., J. EZZELL, B.J. HINNEBUSCH, M. SHIPLEY, E.A. HENCHAL & M.S. IBRAHIM. 1998. A 5' nuclease assay for the detection of *Yersinia pestis*. J. Clin. Microbiol. **36:** 2284–2288.
17. IBRAHIM, M.S., J.J. ESPOSITO, P.B. JAHRLING & R.S. LOFTS. 1997. The potential of 5' nuclease PCR for detecting a single-base polymorphism in Orthopoxvirus. Molec. Cell. Probes **11:** 143–147.

18. IBRAHIM, M.S., R.S. LOFTS, P.B. JAHRLING, E.A. HENCHAL, V.W. WEEDN, M.A. NORTHRUP & P. BELGRADER. 1998. Real-time microchip PCR for detecting single-base differences in viral and human DNA. Anal. Chem. **70:** 2013–2017.
19. NORTHRUP, M.A., B. BEEMAN, R.F. HILLS, D. HADLEY, P. LANDRE & S. LEHEW. 1996. Miniature DNA-based analytical instrument. *In* Analytical Methods and Instrumentation: Special Issue on Micro TAS. H.M. Widmer, Ed.: 153–157. Ciba-Geigy. Basel.
20. BELGRADER, P., W. BENETT, D. HADLEY, G. LONG, R. MARIELLA, JR., F. MILANOVICH, S. NASARABADI, W. NELSON, J. RICHARDS & P. STRATTON. 1998. Rapid pathogen detection using a microchip PCR array instrument. Clin. Chem. **44:** 2191–2194.
21. MILANOVICH, F. 1998. Reducing the threat of biological weapons. Science Tech. Rev. (Lawrence Livermore National Laboratory) June 1998, pp. 4–9.
22. HODGSON, J. 1998. Shrinking DNA diagnostics to fill the markets of the future. Nature Biotech. **16:** 725–727.
23. NORTHRUP, M.A., D. HADLEY, P. LANDRE, S. LEHEW, J. RICHARDS & P. STRATTON. 1998. A miniature DNA-based analytical instrument based on micromachined silicon reaction chambers. Anal. Chem. **70:** 918–922.
24. TOROK, T.J., R.V. TAUZE, R.P. WISE *et al.* 1997. A large community outbreak of salmonellosis caused by intentional contamination of restaurant salad bars. J. Amer. Med. Assoc. **278:** 389–395.
25. KOLAVIC, S.A., A. KIMURA, S.L. SIMONS, L. SLUTSKER, S. BARTH & C.E. HALEY. 1997. An outbreak of *Shigella dysenteriae* type 2 among laboratory workers due to intentional food contamination. J. Amer. Med. Assoc. **278:** 396–399.
26. SIMON, J.D. 1997. Biological terrorism: preparing to meet the threat. J. Amer. Med. Assoc. **278:** 428–430.
27. FRIEDLANDER, A.M. 1997. Anthrax. *In* Medical Aspects of Chemical and Biological Warfare. F.R. Sidell, E.T. Takafuji & D.R. Franz, Eds.: 467–478. Office of the Surgeon General, Department of the Army. Bethesda, MD.
28. LEBEDA, F.J. 1997. Deterrence of biological and chemical warfare: a review of policy options. Milit. Med. **162:** 156–161.

United States of America v. Ray Wallace Mettetal, Jr.

Preliminary Observations

RAY B. FITZGERALD, JR.[a]

Assistant United States Attorney, Western District of Virginia, Charlottesville, Virginia 22901, USA

INTRODUCTION

On August 22, 1995, a man wearing an elaborate but unconvincing disguise was detained by members of the Vanderbilt University Police Department following a citizen's report that he was lurking in a parking deck reserved for faculty and staff of the Vanderbilt University medical school and hospital. Although the officers initially thought the man might have been a patient at a nearby psychiatric ward, they eventually arrested him for trespassing after the man presented obviously false identification documents and claimed that he was on the campus to check up on a woman he thought was seeing someone behind his back which, if true, would have been a violation of Tennessee law.

Following his arrest, a search of his person and his briefcase revealed a large veterinary syringe containing an unidentified fluid, a large-bore "udder infusion" needle used in the dairy industry, a change of clothing, additional disguise elements, and several documents reflecting a detailed itinerary, including notations of which disguise should be worn at which steps of the trip.

Following several days of confusion as to the man's true identity, Vanderbilt and Nashville Metro police eventually identified him as Ray Wallace Mettetal, Jr., a neurologist practicing in Harrisonburg, Virginia. They also determined that Mettetal had resigned abruptly from a neurosurgery training program in 1984. Further investigation revealed that Mettetal had been exceedingly angry with the chairman of neurosurgery at Vanderbilt, Dr. George Allen, whom he blamed for his inability to complete a neurosurgery program, and that the parking area in which he had been seen was the place where Dr. Allen used to park.

Tennessee authorities arrested Mettetal on a charge that he attempted to murder Dr. Allen. (Since that prosecution is pending in the Tennessee courts, nothing further will be said about that case or the evidence in it.) They requested the assistance to the Harrisonburg police in searching Mettetal's house and office in Virginia for evi-

---

[a]Address correspondence to: Ray B. Fitzgerald, 255 West Main Street, Room 104, Charlottesville, Virginia 22901; Telephone: 804-293-4283; Fax: 804-293-4910.
e-mail: rusty.fitzgerald@usdoj.gov

dence of Mettetal's hatred for Allen. Officers of the Harrisonburg Police executed several search warrants in Harrisonburg and searches revealed additional disguise elements, a book on disguise techniques, additional false identification documents, and handwritten notations describing locations in Nashville related to Dr. Allen.

Providentially, an alert Harrisonburg resident read newspaper accounts of the police searches of Mettetal's home and office and recognized Mettetal's alias as a name used by one of his customers who rented a storage unit north of Harrisonburg. (It is interesting to note that, despite having arrested Mettetal and searched his home and office, there were no leads recovered which would have led to the Acorn 6 storage locker absent this citizen's report. Mettetal was very nearly successful in compartmentalizing his true identity and his alias identity.) This resident called the police, who obtained and executed another search warrant at the storage locker. In the locker, they found about six cardboard boxes containing books about the craft of a "hitman," files about poisons and methods of administration, false identification documents, laboratory equipment, and numerous jars of unknown fluid substances. Most chilling perhaps was a group of files reflecting extended surveillance of Dr. Allen's former residence and workplace, including photographs of Dr. Allen's house during several distinct astronomical seasons.

The Harrisonburg officers recognized the potential hazard posed by these unidentified chemicals, especially in light of the bizarre and homicidal reading material with which they were stored, so they called for a hazardous materials response team to secure the evidence. The assistance of the Federal Bureau of Investigation was sought, and Special Agents from the Richmond Division, from the Washington Field office and from the Laboratory responded. Following a long period of analysis, one of the questioned substances was identified as ricin, a biological toxin derived from the seeds of the common castor plant. In the summer of 1996, Mettetal was indicted in Virginia for possession of this toxin for use as a weapon.

LEGAL CONTEXT

The federal charge lodged against Mettetal was a violation of Section 175(a) of Title 18 of the United States Code. That statute provides, in pertinent part:

> Whoever knowingly develops, produces, stockpiles, transfers, acquires, retains, or possesses any biological agent, toxin, or delivery system for use as a weapon...or attempts, threatens, or conspires to do the same, shall be fined under this title or imprisoned for life or any term of years, or both.

This statute was enacted in 1990 for the express purpose of implementing the Biological Weapons Convention ratified by the Senate in 1974. Very few prosecutions have been brought under this statute; it appears that Dr. Mettetal was the third person tried on this charge. The only reported case of a prosecution under this statute is *United States v. Baker*, 98 F.3d 330, (8th Cir. 1996). Messrs. Douglas Allen Baker and Leroy Charles Wheeler were convicted at trial, but Mr. Baker's conviction was reversed and remanded for retrial for reasons having nothing to do with the statute itself.

PROGRESS OF THE CASE

In addition to the usual pretrial motions, Mettetal sought dismissal on the grounds that the statute represented an unconstitutional extension of Congress' power to legislate. That issue was resolved against Mettetal by the District Court, but is likely to be asserted on appeal since there is no appellate authority on this point. Mettetal's defense team originally served a notice that they intended to present a psychiatric defense, which caused a further delay in the proceedings until he could be examined at the Federal Correctional Institute at Butner, North Carolina. When that examination showed that Mettetal was sane at the time of the offense and was competent to stand trial, that theory of defense was abandoned.

The next tentative theory advanced by the defense team was scientifically based. The theory that emerged from the continued discovery requests was that the substance seized had been stored for so long under circumstances that might have caused it to lose its toxic characteristics—to "spoil," in layman's terms. Special Agent Drew Richardson of the FBI laboratory performed some additional tests to confirm that the sample was still toxic, after which the defense abandoned scientifically based defenses and concentrated on disputing the "possessed" and "for use as a weapon" elements of the offense. Indeed, after reviewing Dr. Richardson's qualifications and his scientific conclusions, the defendant entered into a stipulation that the substance seized from the storage locker was, indeed, ricin, and that it was a toxin within the meaning of the statute—an unmistakable illustration of the clarity, accuracy, and credibility of the Bureau laboratories generally and of Dr. Richardson individually.

More than two years after the indictment was returned, trial began in Harrisonburg on July 27, 1998. Jury selection went smoothly despite the rather intense pretrial publicity, and the government presented twenty-two witnesses attesting to various aspects of Mettetal's conduct. The FBI case agent, Joe Harmon, had even located the professor in Mettetal's biology course in premedical school who taught Mettetal about ricin, it source, and its uses. Harmon had also located a classmate who remembered Mettetal's' participation in, and enthusiasm for, this aspect of his premedical study. The defendant's former spouse and his son testified for the United States, as did the probable intended victim, Dr. Allen. It was this kind of detailed and painstaking investigative effort by he FBI and by the local police agencies that provided the abundance of evidence presented at the trial.

On the fourth day of trial, in advance of the proposed testimony by a psychiatrist who had met with Mettetal before the federal case was brought, the court held a hearing on the extent to which this expert would be permitted to testify. The defense offered a Dr. Kenner of Nashville, who proposed to testify that Mettetal suffered from an adjustment disorder with a depressed mood. That evidence was admitted by the Court. Kenner also proposed to testify that, in his opinion, Mettetal did not intend to use the toxin as a weapon, but rather was acting out an elaborate fantasy. This proposed evidence was not allowed, in compliance with Rule 704(b) of the Federal Rules of Evidence, which prohibits an expert from offering an opinion about the ultimate issue in contention.

Following several hours of deliberation, the jury found Mettetal guilty as charged. A presentence investigation was ordered and a sentencing hearing was scheduled for December 2, 1998. Following that hearing, Mettetal will be transported back to Tennessee to stand trial for the attempted murder of Dr. Allen.

SENTENCING ISSUES: ON APPEAL

When the case came for sentencing on December 21, 1998, the statute permitted the Court to impose literally any penalty, from a $20 fine to life in prison for this offense. Unlike most federal criminal statutes which have fairly detailed sentencing guidelines for Courts to follow, there have been too few cases decided for the development of a specific guideline for violations of Section 175. Where no specific guideline exists, Courts are required to analogize this offense to others more common, for which guidelines have been published. For example, in the cases of Douglas Allen Baker and Leroy Charles Wheeler in the District Court for the District of Minnesota, *supra*, that Court reasoned that their conduct in possessing enough ricin to poison 127 people was akin to possession of hand grenades or machine guns or other presumptively aggressive weapons. The district court sentenced Baker and Wheeler each to 33 months imprisonment, 3 years supervised release, and a special assessment of $50.

In the Mettetal case, the United States argued that his conduct was so intricate, lasted so long, and was so finely focused on one identifiable victim that he should be sentenced as though he attempted murder, which would have yielded a sentencing guideline range of 78 to 97 months. On the basis of the quantity of ricin alone, estimated by the FBI as enough for 3,904 lethal doses, Mettetal's sentence should be longer.

The District Court, however, followed a suggestion first put forward by the Probation Service, which likened Mettetal's violation to "Knowing Endangerment Resulting from Mishandling Hazardous Substances" (United States Sentencing Commission, *Guildelines Manual*, Section 2Q1.1), which started the Guidelines computation at 51 to 63 months, which range the Court adjusted upward in consequence of specific offense characteristics, including the volume of the toxin. Finally, the court departed upward from the range eventually computed and imposed a sentence of ten years. Mettetal noted an appeal of the conviction and sentence; the case is awaiting argument in the Court of Appeals for the Fourth Circuit.

LESSONS LEARNED

While this investigation and prosecution must be considered a success by any measure, there were a few times when the agents and I observed instances of "If we had it to do over again…". The first such recognition came about June of 1996, nearly ten months after the execution of the search warrant at Acorn 6, when we first learned that laboratory analysis of the unknown substances revealed 43 grams of ricin. If that examination had been performed more quickly, it is possible that Mettetal's prosecution could have been concluded a year or more before it was. In the meantime and of greater importance, during the time the substances were in the possession of law enforcement but unidentified, there was a greater risk that someone could have been accidentally poisoned. (I understand from Dr. Richardson that his organization now has simple hand-held equipment capable of making a rough determination of the presence or absence of ricin in a sample. Dissemination of this equipment, together with training on their uses and limitations, could dramatically shorten the turn-around time during which the sample remains unknown.)

Similarly, it was fortunate that the Harrisonburg police officers possessed the experience and professionalism to first assess and then react appropriately to this very dangerous crime scene. Not every law enforcement officer would have known that opening and emptying this storage locker was a potentially lethal operation. This points out an opportunity for additional training and liaison with local police agencies about the hazards of participation in this kind of investigation, together with a reiteration of the FBI's expertise and willingness to assist in this area.

CONCLUSION

Prosecutions are collaborative efforts. While it was my privilege to present the evidence at the trial and while I am pleased to offer these observations for those who may be interested, I would be remiss if I failed to emphasize that this prosecution was successful and that a very serious threat to the public was neutralized because of the coordinated efforts of the Vanderbilt University Police Department, the Nashville Metro Police Department, the Tennessee Bureau of Investigation, the Harrisonburg Police Department, the Washington Metropolitan Airport Commission Police Department, the Diplomatic Security Service of the State Department, and the Federal Bureau of Investigation. It is my hope that none of us ever have to face this kind of case again; but I am confident that, if this challenge is posed, these individuals and agencies will respond appropriately and effectively.

Terrorism Overview

PETER S. PROBST

Office of the Assistant Secretary of Defense in Special Operations and Low-Intensity Conflict, The Pentagon, Washington, D.C. 20301, USA

In the past decade, we have experienced a world that is in flux as we moved from the relative stability of a bipolar political model to a period of instability, which threatens to worsen before it ultimately achieves a new and yet to be defined stasis.

The disintegration of the Soviet Union and the collapse of East European communist regimes produced a power vacuum that has enabled violent nationalist, ethnic, and religious forces that were long thought dormant to reassert themselves and contribute to the volatility of the post Cold War era. Violent, militant, Islamic elements—often with the help of state sponsors but also operating independently—now range world-wide and have a demonstrated global reach. Local and regional conflicts, famine, economic disparity, mass movements of refugees, and brutal and corrupt regimes contribute to instability and fuel a frustration and a desperation that increasingly finds expression in acts of terror. Ready access to information and information technologies coupled with the ability to communicate globally via the Internet, fax, and other media provide terrorists new tools for targeting, fund-raising, and dissemination of propaganda, as well as relatively secure and virtually real-time operational communication. Just as the established political order is in a state of fundamental change so is terrorism and thus the nature of the challenge it poses.

Because it is effective and cheap (and sponsorship can be disguised or denied), terrorism increasingly will be a weapon of choice for extremists and rogue states. While terrorism is not new, it is undergoing fundamental and profound change, both in terms of the profiles of its perpetrators and their capability to inflict mass casualties.

In the past, terrorist groups were motivated primarily by political ideology. Even the most brutal groups usually avoided mass casualty attacks for fear of alienating their political constituencies, potential recruits, and other actors able to advance their political agenda.

Whereas politically motivated terrorism appears to be in decline, terrorism carried out in the name of religion is increasingly ascendant. In contrast to their politically motivated counterparts, religious zealots, whether members of a terrorist group or cult, exhibit few self-imposed constraints. They actively seek to maximize the carnage, believing that only by annihilating their enemies can they fulfill the dictates of their guru or god.

The proliferation of weapons of mass destruction and of individuals schooled in their design and construction represent a second development that fundamentally affects the nature of terrorism.

The fragmentation of the former Soviet Union and the lack of adequate controls on biological, chemical, and nuclear stockpiles have resulted in a flood of buyers eager to purchase lethal material from an expanding black market. It also has resulted in a plethora of scientists and technicians, who are either unemployed or underemployed. They have families to feed and are under increasing financial pressure to sell their skills to the highest bidder—whether that be a rogue state, a terrorist group or cult, or an independent operator such as Usama bin Ladin. If you have ever seen your own child go hungry or desperately ill when you cannot afford food, medicine or clinical care, you can better appreciate the stark choices they face. This is a drama being played out in very human terms, and one that impacts directly on our national security interests.

In this "New World Disorder," religious zealotry creates the "will" to carry out mass casualty terrorist attacks; proliferation provides the "means." It is this nexus of will and means that makes combating terrorism a whole new ball game. It is this marriage of will and means that has forever changed the face of terrorism.

This concept is important and much more than an academic distinction of passing interest. Differences in terrorist motivation and mind-set directly affect the terrorist's choice of targets and weapons and, therefore, the motivations of our terrorist adversaries must be of direct concern to all of us.

In a world of competing headlines, terrorists will find it necessary to escalate the carnage in order to maintain their visibility. Publicity is the oxygen of terrorism. As a consequence, we may expect increased experimentation with improvised biological, chemical, and radiological devices as a means to rivet public attention and thereby advance the terrorist's agenda.

Further complicating the mix is what might be called the atomization of terrorism. In the past, the most dangerous terrorist groups were usually politically motivated. They were well organized with a well defined chain of command, and were usually state sponsored or state supported. Today, we are increasingly seeing the emergence of *ad hoc* terrorist groups that are primarily driven by religious zeal. They are relatively amorphous and march to their own and very different drummer.

The most dangerous are those that emerged from the Afghan War waged against the Soviet Union by militant Muslims. From countries as diverse as Pakistan, Spain, Malaysia, Argentina, and the United States, they answered the call. They trained; they fought; and they died in a Holy War to defeat the Russian aggressors. Of those who did not perish, most ultimately returned to their homeland and many of them returned imbued with a fanatical hatred of the West, the United States, and our perceived values and culture.

Their militancy has manifested itself in a commitment to spread their peculiar brand of Islam by violence and terror. They operate transnationally, and the world is their oyster. Ramsi Yousef, the architect of the World Trade Center bombing, is an example of this new form of terrorist zealotry.

Such terrorists do not operate as the pawn of a particular country, although they may exhibit an ideological affinity with states such as Iran and Sudan or even develop *ad hoc* working relationships with the intelligence services of one or more rogue states. Such movements exhibit little or no formal structure and, as in the case of the World Trade Center bombers, their operational cells may be drawn from diverse nationalities in which traditional Shia and Sunni rivalries are increasingly submerged by a shared hatred of the United States and what we stand for.

Much like our own home-grown terrorists, who comprise some of the more violence-prone elements of the so-called Patriot Movement, some of the violent Islamists appear to embrace the principle of "leaderless resistance," coming together in small informal, amorphous groups that operate independently and act when lucrative and vulnerable targets present themselves. With no real chain of command and no higher headquarters giving orders or imposing discipline, they act when they believe it propitious or when some external event galvanizes them to commit yet another heinous act of terror.

Further complicating the picture is that the power to inflict mass casualties has devolved from the super powers to groups and now even to the individual. The ultimate horror may be a biological Unabomber—a Ted Kaczynski armed with pulmonary anthrax, pneumonic plague, or even small pox.

By acting alone, his security is maximized and he need emerge from the shadows only long enough to commit his act of terror and then scuttle back into the fog of anonymity.

I remember talking to my uncle—now long dead—who also was an intelligence officer. I must have been about twelve or so, and he was regaling me with some of his stories of World War II. Normally a jocular man, I remember he suddenly turned deadly serious and grasped my shoulders with both his hands and, in a low, hoarse, almost inaudible whisper, told me he had learned the hard way that once you share a confidence, you must assume it is no longer a secret.

I probed as best a twelve-year old could but never did learn the story, and can only assume it was one of betrayal. He never referred to it again, but it seems to me that the Ted Kaczynskis of the world have taken his dictum to heart, and their ingrained caution and pathological isolation make them incredibly difficult to detect and preempt.

Another relatively new phenomenon is what might be called the privatization of terrorism, with Usama Bin Ladin being the best known example.

One of some 20 sons of a Saudi construction magnate, Bin Ladin in estimated to be worth between $250–$300 million. He uses his wealth to advance his brand of extremist Islam. Establishing a web of business fronts and charitable non-governmental organizations, Bin Ladin has developed a sophisticated terrorist network that spans the Middle East, with representation in several African states, the Central Asian republics of the former Soviet Union, and, even, the United States.

Most recently, he has been charged with bombings of our embassies in Kenya and Tanzania. He is also believed to be working to develop a nuclear capability and has shown an unsettling interest in chemical and biological weapons. Bin Ladin, in fact, functions much like a terrorist state sponsor. Most unsettling to me is that he is not the only financier of terror operating in such a fashion—he is just the best known.

I think probably one of the most significant developments in the facilitation of terrorism has been the Internet. Virtually any terrorist group worth its salt uses the Internet. It provides near real-time operational communication that may be encrypted or run through a series of anonymous remailers and service providers spanning any number of countries and continents to make the identification of the originating site extremely difficult and, at times, impossible. Some groups do not even bother to encrypt their messages but use predetermined coded phrases, seeking to hide their seemingly innocuous messages in the increasing cacophony of this expanding medium.

The Internet, in fact, represents a whole new front in the terrorist war. A colleague at the National Security Agency noted that ultimately he expects to confront terrorist groups that exist solely in cyberspace—members never meet face-to-face except, perhaps, to carry out specific terrorist operations.

The Internet also provides an important tool of psychological reinforcement for the demented, deranged, and those in the society who harbor the most extremist views. Until recently many such individuals thought they were alone and their sense of isolation served to inhibit their acting on their most anti-social impulses. On the Internet, however, they find others who are like-minded and who share their darkest fantasies. Communication can provide the reinforcement that strengthens their resolve and propels them to commit acts of mayhem and terror.

The Internet also serves to enhance terrorist R&D efforts. It is widely used to disseminate information on how to better construct conventional terrorist weapons as well as improvised biological, chemical, and nuclear devices. When I was in school, the big seller among the radical fringe was the *Anarchist's Cookbook*, essentially a "how to" reference. Its contents pale in comparison to what is now available on the Internet.

I heard a story, perhaps apocryphal, that shortly after the Aum Shinrikyo sarin gas attack against the Tokyo subway system, a government contractor posted on the Internet a formulation for sarin that he subtly altered so that the substance produced would be harmless. Within 30 minutes, he reportedly received 50 messages that provided the correct formula and, in exquisite detail, explained why he was in error. Even more troublesome were some of the alternative scenarios for terrorist use of sarin suggested by those who participated in the exchange. Several were truly diabolical and, on examination, were reportedly determined to be scientifically valid.

There is increasing concern that terrorists may target key elements of the national infrastructure. These generally are described as including telecommunications, energy distribution, banking and securities, transportation, military/defense, water supply, emergency services, and public health.

Progress presents us with a curious paradox. As countries modernize, they become increasingly dependent on sophisticated technologies with computers both running and linking vital, once-disparate systems into a national infrastructure. Because of its complexity and interdependence, infrastructure presents a sophisticated adversary with unique targeting opportunities.

Complex national infrastructures are vulnerable because they all have critical nodes or choke points that, if properly attacked, will result in significant disruption or destruction. The attack may be computer generated or rely on more conventional assaults employing truck bombs, dynamite, or cable-cutting to unleash a chain of events in which a service grid, pipeline, or air traffic control system collapses in a cascading effect.

Progress has heightened infrastructure efficiency but the resultant reduction in redundancy has created vulnerabilities that make the infrastructure of the United States an increasingly lucrative target.

Major power failures that black out large parts of the country, systemic problems with the air traffic control system, and breaks in highly vulnerable gas and oil pipelines are covered in detail by the press, discussed on radio talk shows, and dissected and analyzed on the Internet. Terrorists, as part of the attentive public, are increasingly aware that the national infrastructure represents a high value and vulnerable target. Vulnerabilities are of course being addressed, but the process is a slow, arduous, and expensive. Some fall into the "too hard" category.

There are yet other lucrative targets that are of concern to me. One is the nation's blood supply; the other is American agriculture.

We probably all remember the Palestinian terrorist efforts to destroy the export market for Israeli produce by poisoning Haifa oranges with mercury. Or the episode involving the poisoning of small quantity of Chilean grapes with cyanide—an act that crippled the export market for Chilean produce.

It has been my experience that when discussions turn to infrastructure vulnerability, the security of American agriculture and our food supply generally is not addressed. I view this as a serious shortcoming, and one that I expect will be corrected as a result of the issue being spotlighted by this volume and other similar efforts.

United States agriculture, particularly livestock, has been targeted in the past and there is every reason to expect more attacks in the future. Seth Carus, in his excellent volume *Bioterrorism and Biocrimes*,[1] describes the German covert campaign against the United States that took place against the backdrop of the early years of the World War I, when the United States was still neutral. The targets were draft animals that the Allies had purchased in the United States for use in the European war effort and the German plotters carried out an ambitious effort to infect them with glanders and anthrax. (Some 3,500 horses were infected.)

We also know that the Soviet Union during the Cold War had targeted American agriculture as part of its offensive biowarfare program. I am concerned that the Russian scientists and technicians who have worked in such programs, and who are now unemployed or have not been paid in months, may well find irresistible the offers of terrorist groups, rogue states, or Mafia middlemen.

In our 1994 study, *Terror-2000*,[2] we raised the spectre of a malevolent Johnny Appleseed using natural pests and diseases to attack American agriculture. The scenario is chilling, credible, and may be as close as tomorrow's headlines. Agents that could be employed against the American food supply cover a wide spectrum and include wheat rust, foot-and-mouth disease, anthrax, glanders, avian influenza, and African swine fever.

The effect of such attacks on the American economy and the confidence of our people in the government's ability to protect them could provide material for innumerable Tom Clancy novels and nightmares for government officials charged with protecting this country against such terrorist depredations.

I believe there is general agreement that U.S. agriculture, particularly livestock, is highly vulnerable and an attractive target for several reasons: (1) a variety of agents and instructions for their use as a bioweapon are readily available from a multitude of open sources, (2) American crops and livestock are known to be highly vulnerable to these agents, and (3) with the U.S. agricultural industry generating more than $1 trillion in annual economic activity and more than $140 billion in annual exports, a successful attack could well cost the American people billions—providing the terrorists with an extremely high payoff at minimal risk, thereby presenting an almost irresistible temptation to those who wish to do us harm.

Attacking our food supply is not a new idea. The threshold has been breached. Such operations have already occurred. We ignore the issue at our peril.

REFERENCE

1. CARUS, W.S. 1998. Bioterrorism and Biocrimes: The Illicit Use of Biological Agents in the 20th Century. Working Paper. Center for Counterproliferation Research. National Defense University. Washington D.C.
2. CETRON, M. & P. PROBST. 1994. Terror-2000: The future face of terrorism. Office of the Assistant Secretary of Defense for Special Operations and Low Intensity Conflict. The Pentagon. Washington, D.C.

The Changing Biological Warfare Threat
Anti-Crop and Anti-Animal Agents

SHARON A. WATSON[a]

Chemical and Biological Defense Department, SRS Technologies, 500 Discovery Drive, Huntsville, Alabama 35806, USA

At the mention of the words biological warfare (BW), one immediately visualizes a sweeping pandemic in which a disease such as plague spreads rapidly through a population resulting in mass casualties, panic, and death. In recent decades, the spectrum of chemical and biological warfare has expanded to include microorganisms and toxic substances that never could have been weaponized previously. The boundaries between chemical-biological classifications have become blurred and tools routinely used in legitimate industrial settings have the potential of devastating effects in the wrong hands (FIG. 1). Biotechnological advances have opened new doors and made BW capability even easier to acquire. The Internet abounds with information that could be misused by terrorists and hostile developing countries.[1]

Bioterrorism is a very real possibility and its potential for mass destruction is the subject of increasing international concern. At least 17 countries are suspected to have active research and development programs for biological weapons. A handful of those have been implicated as sponsors of international terrorism, and some of those have investigated anti-crop and anti-animal agents. In addition, radical groups and individuals with grievances against the government or society in general have discovered the utility of making chemical/biological threats in their attempts to intimidate, coerce, and disrupt.[2]

Against this backdrop of change in the BW threat, we also have a change in the structure of the world economy and political situation. The pre-eminence of politico-military competition and the bipolar U.S.-U.S.S.R. world is giving way to a new international security environment driven by politico-economic competition. It has been predicted by Bergsten that "The Twenty-First Century will be a century of economic warfare."[3]

The United States is vitally dependent on its agriculture and livestock. We are dependent on plants for our staple crops (wheat, rice, corn, etc.), for fibers (e.g., cotton and flax), for wood, for vegetables, fruits, and luxury items such as tea and tobacco, and for many materials used in industry. All of these economically important plants are subject to attack by plant disease.[4]

Plants are susceptible to a variety of microorganisms. Those featured in various military BW development programs are shown in TABLE 1.[4] The Iraqis admitted to

---

[a]Address correspondence to: Sharon A. Watson, Chemical and Biological Defense Department, SRS Technologies, 500 Discovery Drive, Huntsville, AL 35806; Telephone: 256-971-7018; Fax: 256-971-7067.
e-mail: swatson@stg.srs.com

TABLE 1. Illustrative examples of potential anti-plant agents

Disease	Agent	Note
Viruses		
Tobacco mosaic	Tobacco mosaic virus	Natural airborne transmission
Sugar-beet curly top	Curly top virus	Natural leafhopper transmission
Bacteria		
Rice blight	*Xanthomonas oryzae*	Many modes of transmission
Corn blight	*Pseudomonas albopre-cipitans*	
Fungi		
Late blight of potato	*Phytophthora infestans*	Caused Irish potato famine
Rice blast	*Pyricularia oryzae*	Standardized agent
Black stem rust of cereals	*Puccinia graminis*	*P. graminis tritici* was a standardized agent

the development of anti-crop and anti-animal agents in addition to antipersonnel agents.[1]

Our livestock and poultry industry is also critically important to our economy. Natural disease outbreaks, even those that have been pinpointed quickly and contained in a specific area, can greatly deplete the population and food supplies. For example, in the 1984 outbreak of avian influenza (H_5N_5 strain) in the Eastern Pennsylvania poultry population essentially eliminated the population. Although 40–60%

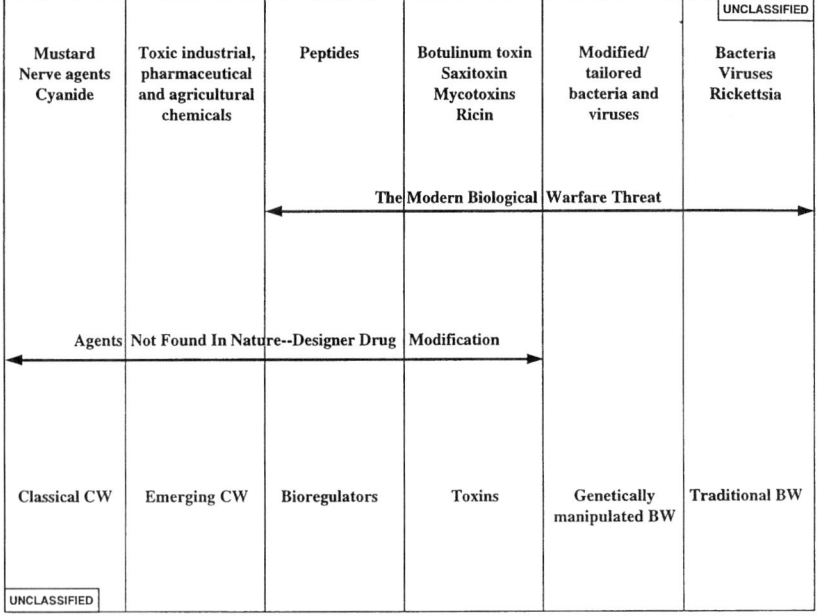

FIGURE 1. Chemical and biological warfare spectrum.

TABLE 2. Illustrative examples of potential anti-animal agents

Disease	Agent	Note
Viruses		
Foot-and-mouth disease of cattle	FMD virus	
Rinderpest or cattle plague		Intensively studied in World War II
Newcastle disease of poultry	NDV	Intensively studied in World War II
Rickettsiae (Bacteria)		
Heart-water of sheep and goats	*Cowdria ruminantium*	Tick carrier
Fungi		
Aspergillosis of poultry	*Aspergillus fumigatus*	

of the birds would have survived the infection, control and containment procedures dictated destruction of all sick and exposed animals.[5]

A covert biological attack could be easily designed to cripple the poultry or livestock industry by simultaneously introducing three or four highly contagious, highly fatal animal diseases. Livestock and poultry populations are concentrated—cattle and swine primarily in the Midwest and poultry in the Southeast (FIG. 2).[6]

Examples of potential anti-animal agents are given in TABLE 2.[4] In an article in *Military Medicine*, Gordon and Bech-Nielsen outlined several potential scenarios to illustrate the potential of such agents. Velogenic Viscerotropic Newcastle Disease (VVND) is an extremely serious disease of poultry and is caused by a virus. The pathology is characterized by severe lesions in the gastrointestinal tracts of the affected birds. In susceptible birds, morbidity rates approach 100% and mortality rates can be as high as 95%. The virus is highly contagious and can be transmitted in a variety of ways including direct contact, contaminated feeding and watering equipment, and by aerosols produced by wheezing, coughing birds. There have been incidents in which flocks were contaminated by movement of contaminated equipment and personnel such as vaccination crews. The most frequent source of natural outbreaks is an infected but asymptomatic bird. Vaccination does not completely protect a flock; a small percentage of birds remain susceptible. To illustrate the consequences of this disease, the authors cite a natural outbreak that occurred in 1971 in Southern California. The virus was probably introduced into the concentrated egg production area by infected smuggled aviary birds. The disease spread rapidly to commercial chicken and turkey farms and mortality approached 100%. A national emergency was declared. The disease was contained and eradicated but the process involved slaughter of nearly 12 million birds and a loss of $56 million.[7]

This incident involved a natural outbreak from a single source, so it is easy to speculate on the damage to the poultry industry of intentional introduction of the virus at a multitude of points in areas of the highest poultry density. Since the aerosol route is such an effective route of exposure, the air handling systems of conventional poultry buildings are a convenient target. An inexpensive dissemination device could be mounted on a motor vehicle or low flying plane and remain virtually undetectable. Alternately, infected birds could be used to spread the disease.

The Newcastle Disease virus is very persistent and disinfection is difficult and expensive. Not only do the infected and exposed birds have to be destroyed, but all ma-

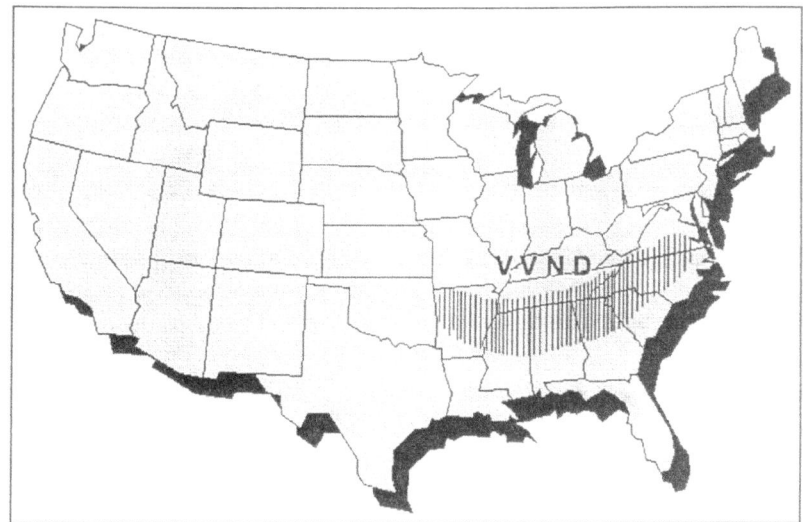

FIGURE 2. Hypothetical scenario.

nure, feathers, and contaminated earth must be removed, incinerated, or buried at least 4-feet deep. Most chicken production buildings have earthen floors. All cages, feeders, watering equipment, floors, walls, and ceilings have to be exhaustively disinfected.

In addition, location of all the infected and exposed birds is extremely difficult because the virus can remain undetected in vaccinated flocks. Humans can also be a factor in spread of the disease. Humans may shed the virus for many days after exposure (up to 14 days).[6]

The U.S. intelligence community has traditionally been focused on foreign antipersonnel biological warfare programs and there are few (or no) intelligence analysts responsible for assessing the BW threat to our agricultural industry. Unfortunately, the responsibility for analysis falls on the periphery of the various agencies' Mission Statements. Every major offensive BW program to date (for the U.S., U.S.S.R., Japan, and Iraq) has developed a variety of agents targeted at non-human populations. If the twenty-first century is indeed the century of "economic warfare," we are extremely vulnerable to this threat. Many agencies and organizations must work collaboratively to ensure that our economy would survive a bioterrorism attack, as well as to ensure that such activity would be recognized and distinguished from a natural occurrence. Educational tools, databases, and training programs need to be rapidly developed in the near-term to increase awareness and ensure national preparedness against bioterrorism attacks of all varieties.[1]

REFERENCES

1. WATSON, S.A. 1998. The Biological Warfare Threat. Secret Level Briefing. SRS Technologies.

2. COHEN, W. 1997. Proliferation: Threat and Response. Office of the Secretary of Defense DTIC ADA332435, Washington, D.C.
3. BERGSTEN, C.F. 1994. Japan and the United States in the New World Economy. The United States in the World Political Economy. McGraw-Hill. New York.
4. DANDO, M. 1994. Biological Warfare in the 21st Century. Macmillan Publishing Company. Riverside, NJ
5. FOREIGN ANIMAL DISEASE REPORT. Veterinary Sciences. 1984. U.S. Department of Agriculture. Emergency Programs **12**(2): June.
6. COMMITTEE ON FOREIGN ANIMAL DISEASES OF THE UNITED STATES ANIMAL HEALTH ASSOCIATION. 1984. Foreign Animal Diseases—Their Prevention, Diagnosis, and Control.
7. GORDON, J.C. & S. BECH-NIELSEN. 1986. Biological Terrorism: A Direct Threat to Our Livestock Industry. Military Med. **151:** 357–363.

Trends in American Agriculture
Their Implications for Biological Warfare against Crop and Animal Resources

WALLACE A. DEEN

U.S. Army, Retired, 39337 West 120th Street, Richmond, Missouri 64085, USA

INTRODUCTION

Much of the actual thought on biological warfare against crop and animal resources was a product of the era of 1950–1965. Practically all of the literature on this topic was produced during this same period. Much has changed in American and world agriculture since that time. These changes have altered the nature of the agriculture target and thus the entire scope of biological operations in this arena.

Historically, anti-plant and anti-animal agents were selected for widespread distribution, in a wartime situation, with the intent of killing or rendering unfit for their intended use. For example, during World War II anthrax was pressed into tons of oilseed cake by the British from December 1942 to April 1943. When dropped and consumed, the resulting disease would deny the Germans access to horse-drawn transportation and destroy cattle, sheep, and swine which served as both food, raw materials for clothing, and a chemical feedstock for the munitions industry (animal fat was used to produce nitroglycerin).

Biological warfare threats still encompass denial of food supplies, but now includes economic objectives, primarily economic loss to the industry by restrictions on international trade and disruption of internal distribution by governmental efforts to isolate and eradicate the disease.

MAJOR TRENDS IN AGRICULTURE

Many of us still retain the idea of the family farm of the 1940s and 1950s, a mostly self-sufficient venture supported by a diversity of animal and plant life. We still see the remains of the old farmsteads with chicken houses, pig pens, and barns for cattle and storage of a variety of grains. Some of this exists in the part-time farmers on the urban edge, but sustainable agricultural production is different.

Today's agriculture is shaped by powerful trends: concentration, decreasing genetic diversity, consolidation of support industries, urbanization, and internationalization of trade. These trends have changed agriculture completely. The biological agents developed by either countries, terrorists, or individuals have also changed.

Concentration

Concentration has occurred on individual farms and within regions. Improved transportation and the economic necessity to lower the cost per unit profit of agricultural commodities have accelerated this trend.

Individual farms are larger and concentrate on the production of large volumes of a single commodity group. The earliest changes were noted in the poultry industry, changing from many small "backyard" flocks to large commercial operations of 250,000 to multi-million bird sites. Typical poultry operations produce about a $.04 per pound margin to pay for labor. To produce a near-poverty level income of $16,000 requires the operation to produce approximately 4,000,000 pounds of poultry. Swine and beef production is also consolidated. According to industry officials, the top 40 pork producers control 36% of production today, and, by the year 2002, the top 40 producers will control 90% of the production. In the beef industry, by 2010 the largest 30 feeders will generate 50% of the finished cattle.[1]

The second impact of concentration is geographic. Improved transportation has untied the point of production from its market. Specialty products are relocating to areas that offer geographic, climatic, regulatory, or financial advantages. This has resulted in concentrations of certain animals in specific areas, such as poultry in Pennsylvania, Maryland (eastern shore), Virginia, Georgia, and Arkansas; swine in North Carolina and Iowa; and cattle feeding in western Kansas and the border areas of surrounding states.[2]

For the user of biological agents, the trend to concentration has reduced the target's geographic area, increased the potential for spread of infectious agents, and magnified the impact of limited use. On the other hand, concentration allows the defense to concentrate its resources for detection, prevention, and restoration.

Decreasing Genetic Diversity

Closely associated with the increasing scale of agriculture is a decreasing genetic diversity. Unlike concentration, decreasing diversity is common to both plant and animal agriculture and is driven by the combination of the buyer's requirement for uniformity of product and the producer's desire for maximum productivity. In the animal area, we have seen the virtual disappearance of the brown egg because the breeds that produce brown eggs are heavier and thus more costly to maintain. In the poultry and swine industry, reduced diversity produces less variability in the product and facilitates automation in preparing the end product for the consumer.

The advent of commercially viable biotechnology products in the plant kingdom has even more radically reduced the diversity of the commodities. The desire for superior yields and quality coupled with current gene technology have narrowed the genetic base. Most dramatically, a recent ability to transmit resistance to certain chemicals (herbicides) has resulted in a near monoculture in large plantings of soybean and corn in the United States.[3]

Decreased genetic diversity is a benefit for the well-financed, sophisticated user of biological agents. The lack of diversity presents the unique opportunity to create a designer organism tailored to the known genetic material of the target. The universal vulnerability of the target also adds to the difficulty of separating a natural infection from an induced disease.

Consolidation of Support Industries

Followers of the stock market have certainly noted the consolidation of suppliers and consumers of agricultural commodities. An outstanding example is Monsanto, which has made the transition from a chemical to agricultural biotechnology company. This trend, driven by the economies of scale, has reinforced the earlier trends of concentration and lack of genetic diversity by reducing the number of seed stock producers and limiting the sources of supply.

For the biological agent user, this consolidation offers a viable additional method of dissemination. The consolidation reduces the effort and resources required to contaminate a product. A single incident could produce much more damage than in a more dispersed source. From the defensive, less resources would be required to focus on preventive measures.

Urbanization

The trend of urbanization has dominated the last portion of the twentieth century. Agriculture has been virtually eliminated from the fertile river valleys and has been pushed away from the population centers. This trend has been accelerated in recent years by concern for aesthetic and perceived hazards of agriculture arising from disposal of animal waste, use of chemicals on plants, and avoidance of dust and odor. Less than 3% of the U.S. population is considered rural and less than 1% of the population is dependent on agriculture for a livelihood. As a result, the average city has less than a 5-day food supply and that food must travel more than 1,300 miles from field to table.[4]

Urbanization is beneficial for the agricultural producers because it isolates them from the major sources of international movement of plant and animal diseases. Conversely, the urban areas are further removed from their sources of supply. Disruptions in food supplies are much more likely to be felt in a short time and restoration requires a longer period of time.

For the user of biological agents, the urban areas are becoming more vulnerable to disruption. Concentration of population increases the pressure applied to governments to alleviate the situation. The defense is reduced to very costly procedures to assure nutrition, but not satisfactory food, to the population of the urban area.

Internationalization of Trade

This is the most important development for those who would use biological agents against plants or animals. Effective use no longer requires that massive and widespread devastation occurs; the mere presence or suggestion of presence of an organism can have economically devastating consequences. Internationally, this is well illustrated by the number of "free trade" pacts that retain existence of an animal or plant disease as the single most important restriction in international trade of plant and animal products. Even the most casual observer can recall the continuing turmoil over bovine spongiform encephalopathy (BSE) and its effect on the British beef industry.

A biological agent user can achieve economic effects with the mere presence of a disease or create a lesser level of disruption by alleging that such a disease exists. It reduces the resources required for successful biological operations to a level achievable by a single individual.

SUMMARY

Current trends in American agriculture have changed the vulnerability to use of biological weapons against plant and animal resources. The major effect has been a requirement to look again at the model of the U.S. BW program of widespread dissemination of agent and look to attack models requiring much lower levels of resources. The U.S. biological warfare program models must take the effects of these major trends into account when considering the possible widespread dissemination of a biological agent. The models must also acknowledge the lowered levels of resources required to make such attacks given the modern trends in American agriculture.

REFERENCES

1. High Plains Journal. 1998. August 17: 23-B.
2. The U.S. Livestock Market for Veterinary Medical Services and Products. 1995. American Veterinary Medical Association.
3. 1997. Science News **152:** 104.
4. LIND, L. 1997. Comments of a Michigan State University cultural anthropologist at a seminar at the University of Missouri Agricultural Science Week.

Agroterrorism

Agricultural Infrastructure Vulnerability

JURGEN VON BREDOW,[a,e] MICHAEL MYERS,[a] DAVID WAGNER,[a] JAMES J. VALDES,[b] LARRY LOOMIS,[c] AND KAVEH ZAMANI[d]

[a]*Office of Research, Food and Drug Administration, Center for Veterinary Medicine, Laurel, Maryland 20708, USA*

[b]*U.S. Army Chemical and Biological Defense, Edgewood Research, Development & Engineering Center, Aberdeen Proving Ground, Maryland 21100, USA*

[c]*Development and Engineering, New Horizons Diagnostics Corporation, 9110 Red Branch Road, Columbia, Maryland 21045, USA*

[d]*Office of Director of Defense Research and Engineering, Pentagon, Room 3D375, Washington, D.C. 20301, USA*

ABSTRACT: The intentional contamination of animal feed to reduce the availability of animal-derived human food or to infect human populations is seldom mentioned, but animal feed could be an easy target for bioterrorists. The period of delay between the contamination of the animal feed and adulteration of the human food product provides an additional degree of uncertainty about the source of the contamination and minimizes the possibility of apprehending the terrorist. The less obvious and more natural the source of biological contamination, the greater the likelihood that the animal feed contamination will be mistaken as a natural phenomenon. However, the problems related to managing natural food contamination and intentional food contamination remain the same. Rapid testing and separation of contaminated feed are important steps, followed by the more specific identification of the contaminant to determine the source of adulteration and/or the possibility of decontamination. At this time identification of the bioagents is dependent on the availability of antibody-specific test systems. The rapid development of specific antibodies for the development of sensitive and specific test kits is the key to identifying contamination and dealing effectively with the disposal or decontamination of the animal feed and, ultimately, preventing the contamination of animal-derived human food products.

INTRODUCTION

The contamination of human food and the subsequent reduction of food available for humans is often discussed in bioterrorism conferences. However, the intentional contamination of animal feed as a method of indirectly affecting animal-derived human food or to infect human populations is seldom mentioned. One goal of the terrorism is to create fear in the population. Causing concern about the most fundamental requirements of life will reach the greatest number of people and will

---

[e]Address correspondence to: Jurgen von Bredow, Office of Research, FDA–Center for Veterinary Medicine, 8401 Muirkirk Road, Laurel, MD 20708.

achieve a goal of political unrest. Contamination of water supplies is understood to be a potential bioterrorist objective and thus water supplies of major cities are usually well protected.

The use of biological agents as weapons to inflict a range of political and disruptive effects has been well reviewed by W. Seth Carus[1] and he has concluded "[c]ontamination of food has resulted in the largest numbers of casualties from the use of biological agents." Our goal is to evaluate the potential of affecting animal-derived human food through the contamination of animal feed with bioagents.

It would seem evident that the approach of a terrorist would be to use a very toxic biological agent in the animal feed sufficient to infect or kill a great number of animals in order to make a statement and leave a lasting impression. In New York, a suspicious death of a food-producing animal will lead to a contact with Dr. Larry Thompson,[2] of Cornell University and the New York State Diagnostic Laboratories. This laboratory will autopsy the animal to determine the cause of death and will decide if feed may have been involved. If the feed has been contaminated, an attempt will be made to determine if the contamination is due to a natural source or to a deliberate act of adulteration. A deliberate incident, using a very toxic agent, will be detected in a short period of time by the state diagnostic laboratories and the animals will never go to slaughter. The food animal may die, but, the animal will not become a source of animal-derived human food.

Dr. Larry Thompson and other contributors to this volume suggest that the use of a biological agent that induces no noticeable effect in the live food animal will create a greater hazard. The apparently normal animals will be slaughtered to produce contaminated food.[3] This will cause unexplained outbreaks of food poisoning in various areas of the country for which no one seems to be responsible.

Dr. Carus has suggested that the target would be limited to foods that are commonly eaten uncooked or to rely on a spore or toxin that will survive cooking. But heat-sensitive bacteria, which contaminate human food, cause many cases of food poisoning. The organisms responsible for contamination leading to food poisoning include *Salmonella*, *Staphylococcus aureus*, *Clostridium botulinum*, *Clostridium perfringens*, *Escherichia coli* 0157:H7, and *Shigella*. Even a normal amount of food contamination leads to the loss of significant amounts of food annually. In July of 1998 several positive tests for *E. coli* and a possible related illness led to the recall of 170,000 pounds of ground beef by Costco.[4]

Contamination of normal food can be increased by using feed contaminated with these same bacteria, it should be possible to induce a greater incidence of food poisoning and/or increased costs to the producer and consumer. The normal incidence of bacteria in steers and heifers[5] is 4.2% *Staphylococcus aureus*, 4.1% *Listeria monocytogenes*, 4.0% *Campylobacter jejuni*, 2.6% *Clostridium perfringens*, 1.0% *Salmonella*, and 0.2% *E. coli* 0157:H7. A good choice for undetected contamination of food is a bacterial agent that can pass through the cow or beef to contaminate the meat without significantly affecting the health of, and drawing attention to the condition of, the adulterated animal. *Salmonella* and *E. coli* are natural contaminants in beef and a very low baseline level of these bacteria can be isolated from carcasses at slaughter.[6] A low level of bacterial contamination is accepted as unavoidable and is not considered to be a hazard unless the level of bacteria in a product is greater than 1,000,000 organisms/gram. A level of one million organisms per gram is not normally achieved unless the meat is left at elevated temperatures for prolonged periods before or after cooking. The exact level required to cause illness has not been defined and may vary with the health of the individual ingesting the food product. Since reliable decontamination of contaminated ground beef may not really be possible, the presence of pathogens in any amount will require the elimination of this food product.

Compared to human food, animal feed is one of the least guarded sources of the food supply. Due to the vast amounts of feed stuff required to feed poultry and livestock, it is hardly possible to secure all of the feed. Only a small amount of contaminated feed would produce a significant effect. A test indicating the positive presence of contamination in the feed or in an animal-food product may occur without producing a single human casualty, but the positive test result will create an alarm about contaminated food. The positive tests for *E. coli* in the Hudson meat plant caused millions of pounds of ground beef to be condemned.

It is important to realize that cattle feed typically consists of three main components: hay (which is usually grown on the farm), silage is made from the hay or from corn (which may also be grown on the same farm), and dry cereal grains, corn or more complex pelleted feeds (which are developed by a central feed mill and are shipped to the farm). Only the dry grains and pellets from the mill could serve as an adequate concentrated source to be contaminated at the mill or on the way to the farm.

The intentional contamination of human food with bacteria is not difficult and has already been accomplished. In 1984, salmonella was added to lettuce in a salad bar and resulted in the illness of numerous persons as described in the review by Carus. Carus also describes numerous cases of the employment of bacteria to cause a devastating effect. The general procedure of obtaining bacteria and using them to contaminate feed may occur in the following order: (1) obtain a source of bacteria, (2) add broth and allow multiplication, (3) place broth in flat pan and allow to air dry, (4) scrape dried medium, (5) mix with limestone or other feed additive, (6) spread contents into feed-mill hoppers.

The addition of the final product to the feed is not a simple tossing of the bacteria into a hopper of feed. A terrorist must add the bacteria at a point in the manufacture process that is not to be subjected to heating. Some knowledge of feed mill operations is required to choose a point at which the bioagent will survive and become a part of the feed. FIGURE 1 diagrams potential points where the bioagent may not be added without significant loss due to heat.[7] Even in the steps that apply heat, some of the added bacteria will survive to increase the potential for food contamination. The contents of some large mixing chambers are constantly sprayed with additives from stainless steel tanks. The addition of bacteria to one of these tanks has the potential for contaminating a larger quantity of prepared feed. A final product sprayer may add contaminants as a final step that is no longer subject to heat treatment. The addition of the bacteria to the calcium carbonate used in many feeds is a possibility that could be effective and unique in its simplicity.

Since the total amount of feed manufactured exceeds 110 million tons annually,[8] only a small amount of feed can actually be intentionally contaminated. The chance of finding the contaminated feed would be difficult, since it would have been fed and none remains available for analysis. This situation leaves the investigators with a difficult problem. The evidence is a significant increase in the amount of contaminated

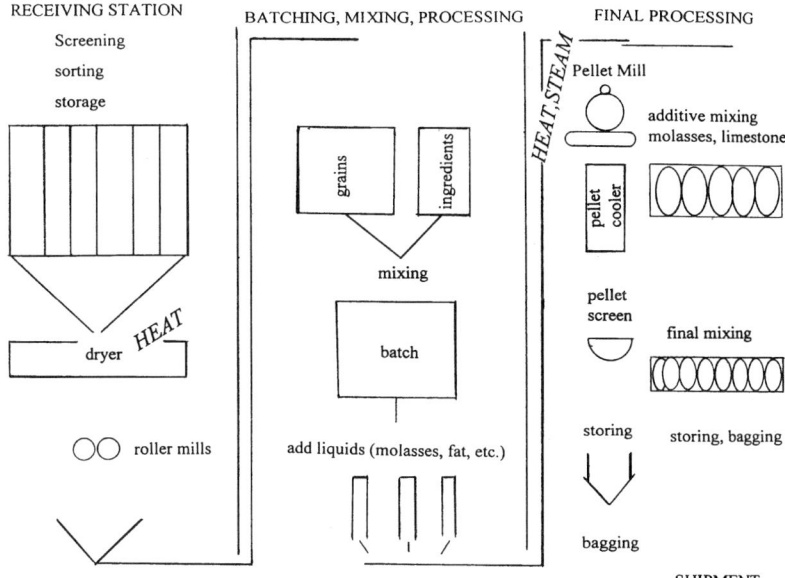

FIGURE 1. Diagram of a typical feed processing plant.

meat without an obvious source of contamination. This will leave the animal feed as a possible source of the problem. Even the lack of evidence may not completely prove that contaminated feed did not exist to cause the contamination of the final human food product.

This presentation could end here, as many others have, with only the indication of the possibility of the contamination of animal feed and its consequences. However, if contamination is a possibility, then proposals should be submitted for dealing with this potential threat. Most important is the fact that animals must be fed and a decision about what to feed them must be made in a short time.

As Franz[9] suggested in his presentation, the first step is the identification of the contamination. In the case of animal feed, the first step should be the identification of feed that has not been contaminated and thus may be used to continue to feed the large population of animals. Since the amount of feed required is enormous and only a very small fraction could be contaminated at any specific time, the most effective tests will identify the lack of contamination and thus prevent the feeding of contaminated feed. Since most of the feed will remain uncontaminated, but all of the feed must be tested to be sure, a test procedure for the remaining feed must be simple, rapid, and provide a clear indication of the condition of the feed. Most tests will be conducted just to prove the absence of bioagent in each lot of feed. This concept is not as unworkable as it may seem, since it is applied to the milk industry. Every truck-tanker load of milk must be tested for beta-lactam drug residue contamination and of the total of 4 million tankers tested, only one in ten thousand actually contains drug residue.[10] The screening tests used for this purpose are reliable, rapid, and in-

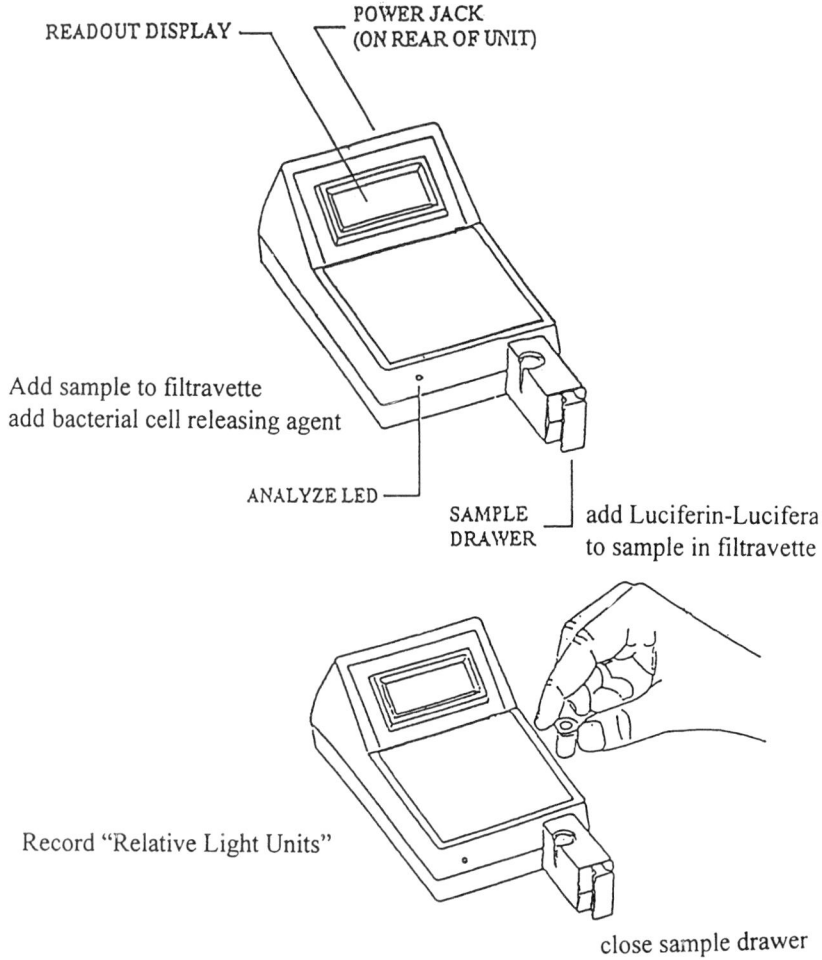

FIGURE 2. Operation of the luminometer.

expensive and the test procedure is a normal part of interstate milk shipping and processing.

Currently one of the most rapid methods for the large-scale examination of animal carcasses for the presence of live bacteria is the luminometer.[11] A diagram of the model developed by the New Horizon Diagnostics Corporation is shown in FIGURE 2. The luminometer is simple, small, and portable and is currently being used on-site to determine the level of bacteria in poultry or beef carcasses. Contamination of the carcass is the result of fecal contamination, which is unavoidable during the slaughter process. The luminometer determines the total of all types of live bacteria in the samples being examined. The utility of the luminometer is to provide a rapid estimate of the total number of bacteria on the carcass, and the instrument is sensitive to levels of bacteria as low as one thousand organisms. The same type of estimate may be possible for bacteria in samples of feed.

The principle of operation of the luminometer is based on the luciferin/luciferase activity. Adenosine triphosphate is released from live bacterial cells and is allowed to interact with luciferin and luciferase to produce light activity, which is indicated on the meter of the instrument. A direct relationship exists between the amount of light emitted as recorded on the instrument and the number of live bacteria in the biological sample.

Bacterial contamination of a sample of feed will be detected as a significant increase in luminometer activity above the normal background. The increase in luminometer activity beyond background will indicate that the feed should not be fed and should be saved for further evaluation. In contrast, a luminometer reading near background would indicate the absence of significant bacterial contamination, an assurance that the feed is safe to use. A preliminary experiment was performed using a small sample of whole oats, which is a basic ingredient of many prepared feeds. The background activity on the luminometer scale indicated a reading of 110. When the same sample of oats was contaminated with one million colony forming units of *E. coli*, the luminometer reading increased to 65,000! This very high activity reading indicates the presence of bacteria and the warning "do not feed." The luminometer is available today and is used to determine the level of bacteria on the surface of carcasses, but it has not been validated for the monitoring of bacteria levels in animal feed.

Other research approaches to monitoring the level of bacteria in feed may also be acceptable. A proposal has been submitted by Mark Seaver of the Naval Research Lab for an optical classification of biological agents through the use of a three-dimensional light scattering and fluorescence technique to determine viability of bacteria.[12] The system may provide an even more rapid method of surveying feed for indications of contamination.

These basic procedures are valuable in answering the important initial question "Is the feed safe to use?" The contaminated feed should be separated and saved for the specific identification of the nature of the agent adulterating the feed. Identification of the specific agent in the contaminated lots of feed will characterize the toxic hazard, validate the results of the initial screening test, and provide a more accurate estimate of contamination. Identification of the specific biological agent may be useful in defining decontamination procedures. Perhaps simple heating of the feed will be sufficient to provide decontamination.

Numerous diagnostic systems for the determination of biological agents already exist, but, they need to be reviewed for the ability to monitor contamination in animal feed. Many diagnostic systems are too slow to provide a rapid definition of the contamination. One efficient system for the identification of agents is the SMART (sensitive membrane antigen rapid test) system developed by New Horizons Diagnostics Corporation.[13] This screening test system is flexible and can be adapted to identify many different types of antigens including: anthrax, brucella, botulism toxin, staph enterotoxin B, tularemia, and ricin. The basic concept and use of the SMART system is shown in FIGURE 3. The system may be modified for additional biological agents in a relatively short time as soon as the appropriate antibody is available.

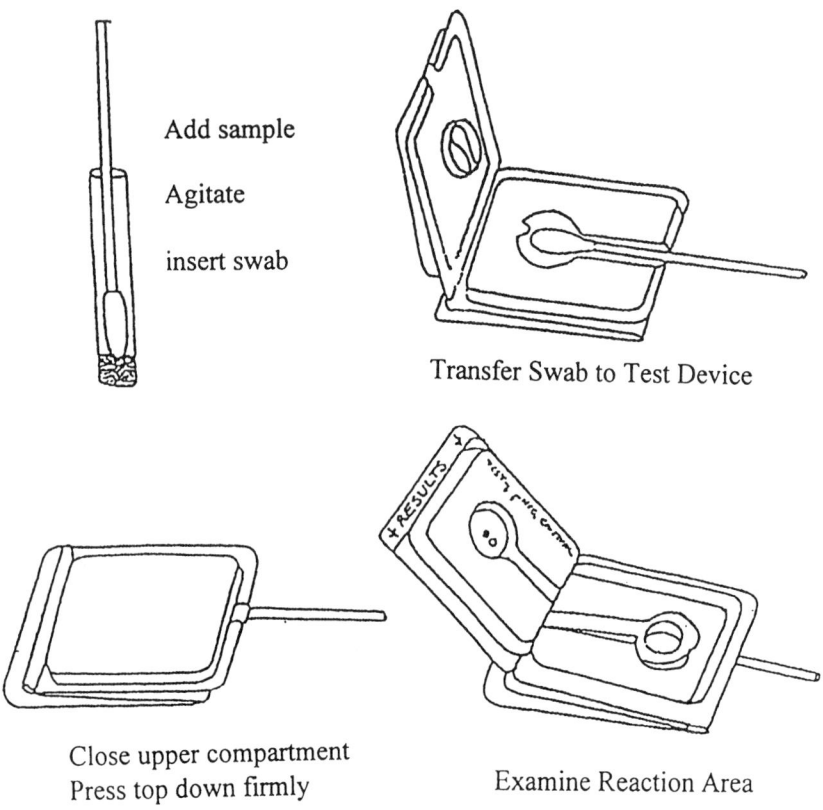

FIGURE 3. Operation of the SMART system.

The antibody is the biological recognition site of the assay system, which defines the specificity and the sensitivity of a test system for antigens (biological agents, drugs and drug residues, or substances to be assayed with the antibody system). The diagram in FIGURE 4 illustrates the combination of colloidal gold with an antibody

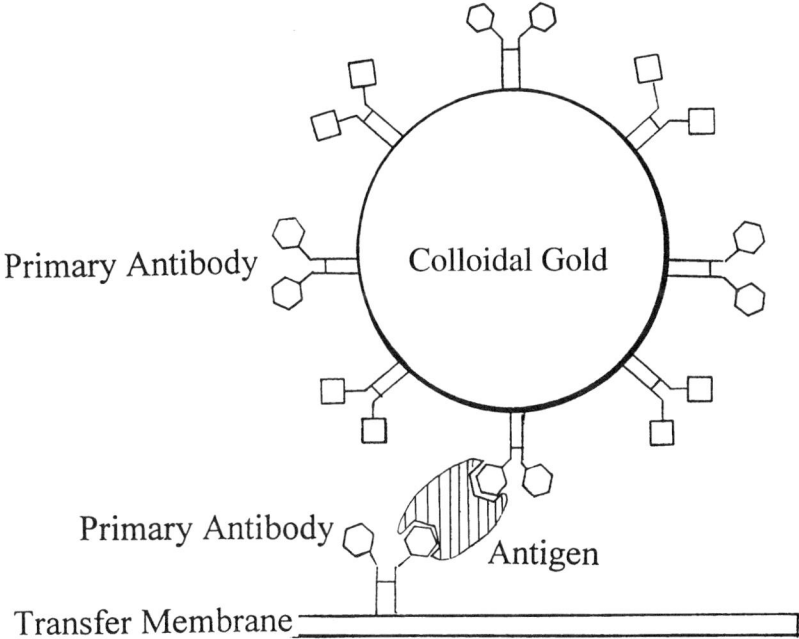

FIGURE 4. Antibody coupled to colloidal gold.

specific for an antigen or a class of antigens. The antigen attached to the antibody-colloidal gold is also attached to another antibody mounted as a spot on a nitrocellulose matrix. The antigen attaches the colloidal gold to the spot. The more antigen present in the sample, the more colloidal gold is attached, making the spot more dense and more visible. The process of attachment to form a visible spot is illustrated in a three-dimensional form in FIGURE 5.

The SMART system is a two-step process requiring a mixing of the sample with the reagents and then placing the resulting mixture on the SMART system, leading to the formation of a spot if the antigen is in the biological sample. The entire process can be simplified to a one-step procedure in which a small quantity of the sample is placed directly onto a flat strip diffusion matrix. The sample containing the antigen is added to the reagents attached to a strip which allows the antigen to mix with the colloidal gold–antibody and form a stable complex.

The complex migrates forward in the flat strip diffusion matrix by capillary action to the antibody stripe. If the antigen is present in the complex, it will bind to the antibody stripe resulting in the formation of a visible line of dense colloidal gold. The lateral flow concept is shown in FIGURE 6 and is called ALERT (antibody based lateral flow economical recognition ticket).[14] The principle of the operation of the ALERT system is described in FIGURE 7.

The key to making both the SMART and ALERT system function in a sensitive and specific manner is the antibody. The development of specific antibodies for these

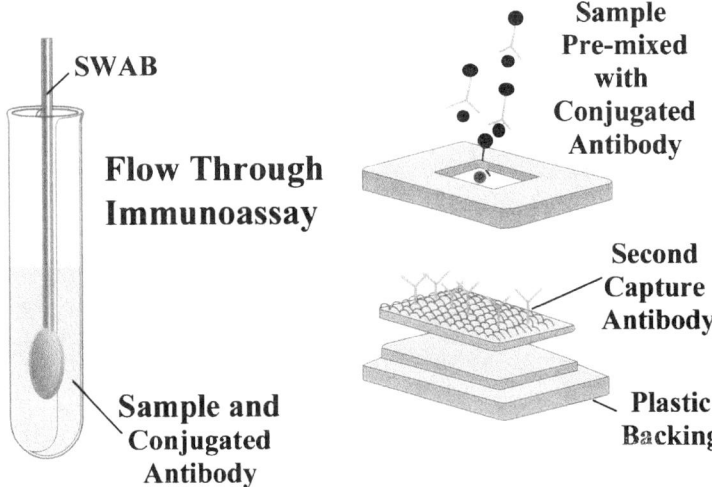

FIGURE 5. Mechanism of spot formation using colloidal gold–coupled antibody.

test systems is usually the most difficult and time-consuming aspect of the test kit. The development of an antibody to an antigen, or the compound to be assayed, requires a well-defined sample of the compound. The antigen is developed into a protein complex that is injected into animals to stimulate production of an undefined mixture of antibodies (polyclonal antibodies) in the sera of the individual animal. Polyclonal antibodies are easy to produce in relatively large amounts and with the proper antisera can result in a useful test kit. However, by the very nature of biological variability, not all animals produce the same spectrum of antibodies when challenged with the same antigen. Production of antisera can require relative large amounts of antigenic material to produce a satisfactory antisera. Typical production of antisera requires immunization of several animals to produce one or two animals producing antisera with the desired qualities. Polyclonal antisera are essentially an undefined mixture comprising antibodies of various immunoglobulin classes, each of which has a different binding affinity. There may also be multiple binding sites for a particular antigen.

Greater characterization, identification, and lot-to-lot consistency of antibodies for the test kits can be obtained through the use of monoclonal antibodies. Monoclonal antibodies are made by fusing splenic B lymphocytes from immunized mice with a myeloma, or cancer cell. This fused cell, or hybridoma, has the immortality of the myeloma cell and produces antibody. Large-scale antibody production is initiated either by production of ascites fluid or through *in vitro* culture of the cells. Ascites containing the monoclonal antibody is produced by intraperitoneal administration of the monoclonal antibody–producing cells into specially prepared mice and collecting the resulting fluid. Alternatively, the cells can be grown in a culture, and the spent tissue culture media that contains the antibodies is collected. While monoclonal antibodies have many advantages over polyclonal antisera, there still is the uncertainty of producing an adequate antibody at the end of this process;

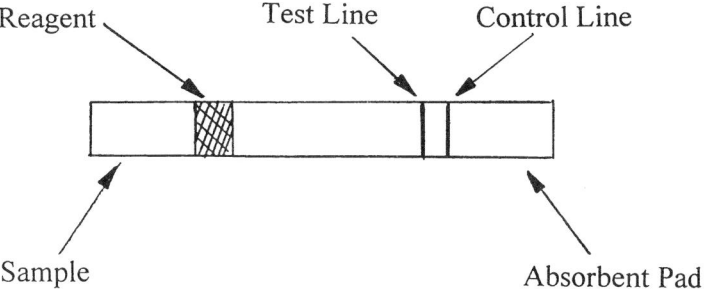

FIGURE 6. Design of lateral flow device.

several cycles of producing new clones may be required to obtain one useful cell line. Again, as with polyclonal antibodies, the probability of success is diminished by the inherent variability of the starting biological systems used to produce the antibodies.

An alternative approach is the production of bacteriophage libraries that contain all the genes coding for antibodies. This method couples the genes encoding a given antibody to one of the bacteriophage coat proteins. This process eventually results in a phage that expresses the antibody on its surface. A phage with reactivity to a particular antigen can be obtained by affinity selection through a process of biopanning. These individual bacteriophage are amplified many times by selection steps and then used to produce antibodies in large-scale bacterial fermentators.

As bacterial cultures are relatively inexpensive, the resulting antibodies are more economical to produce than polyclonal antisera or monoclonal antibodies. In addition, the biological uncertainty of rapidly obtaining suitable antibodies can be eliminated by creating a bacteriophage library containing almost all possible antibody specificities. The existence of the phage library allows for very rapid selection of several suitable clones in days instead of weeks or months. The selection process and rapid production of a stable antibody may make the entire process of recognizing the presence of a new threat agent to the development of a screening system take less than one month.

Any new test system must be validated to function correctly every time that it is used. The evaluation should be carried out in an independent laboratory in the presence of well-defined positive and negative control standards. Validation studies are required to determine if a test kit will reproducibly provide a positive response to a level of contamination in the feed every time that a test is used (i.e., no false-negative response). Equally important is the fact that the test kit must provide a lack of a positive response (i.e., no false-positive response) when the contamination is not present in the feed sample.[15] The absence of a response may be accepted as the absence of contamination with one test kit designed to respond to a specific agent. A positive response to a biological sample may be the result of non-specific cross-reactivity with the antibody, therefore, a positive screening test response must be confirmed with another test kit that uses a different mechanism of action or a different type of antibody likely to cross-react in the same manner.

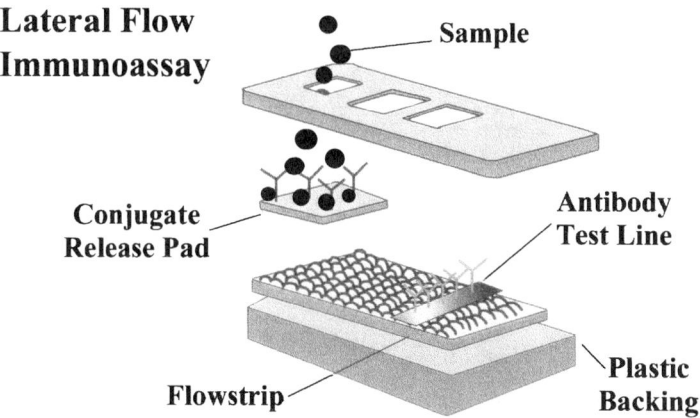

FIGURE 7. Three-dimensional view of lateral flow immunoassay system.

The test kit must be evaluated in the presence of each type of feed matrix that will be encountered. Standard operating procedures must be prepared for the use of each test kit and each type of feed matrix. Standard feed samples must be prepared that are uniformly contaminated with a known level of well-characterized bioagent[16] to function as positive and negative standards to be used during the evaluation of a feed sample. This will validate the fact that the test kit was functioning correctly at the time that the test was performed. Other potential contaminants must be identified to determine if potential cross-reactions may take place that could lead to a false-positive response. The full range of laboratory conditions required to make the test function must be well defined. Test kits that utilize a visual color end-point, instead of an instrumental response, should be evaluated by a panel of individuals to be sure that the color indication can be recognized by all persons performing the test. Training in the use of the tests and all factors affecting the function of the test system should be provided to be certain that the correct interpretation of the test kit response will be made. The contaminated feed samples must be preserved in a manner that will allow additional testing to allow complete characterization of the toxic substance.

One important advantage of specific identification of the contaminant is the possibility of degrading the substance. As some presenters have suggested in this volume, if the contaminant is a biological substance, it may be sensitive to heat or to UV light. Experiments have demonstrated that short wave ultraviolet light energy will kill bacteria. This form of energy may be supplied by UV mercury vapor lamps, which function at a near optimum of 260 nanometers.[17] A major advantage of this technique is that nothing is added to the feed, it does not alter the physical form of the feed, nor does it affect the taste properties of the feed. It may be possible to radiate the feed components at any stage in the manufacture process to include the final feed product before it is placed into bags of feed. If the process is simple and inexpensive, it may become a normal phase in the feed production process to eliminate the potential of feed contamination with a bacterial agent. This would not prevent all avenues of feed contamination but it could eliminate a major route of adulteration.

An alternative to the continuous wave UV light is another waterless, non-toxic surface decontamination/sterilization system using a wide-spectrum xenon flashlamp. This system not only kills pathogens but removes the contaminants through a process called photo-ablation/photo levitation. This photo-optical process has been successfully tested in environmental clean-up of contaminated surfaces. The process is thought to be based on the principles of quantum physics (rather than chemistry), where a photon flux from a wide-spectrum xenon flashlamp (or a laser source) photo-dissociates surface contaminants, which are then taken away using vacuum or flow of an inert gas.

The nature of the contaminant and the surface to be cleaned determine the power and duration of the pulse of light. The process (1) eliminates the need for water in the decontamination process, (2) facilitates destruction and removal of a variety of agents in a single step, and (3) may be utilized for animal feed, vegetables, and fruits.

In preliminary experiments at the Walter Reed Army Institute of Research and through a Cooperative Research and Development Agreement (CRADA) with a flashlamp manufacturer and collaboration with USAMRIID, the flashlamp technique proved successful in killing *Bacillus globigii* spores (an anthrax simulant) in one millisecond. In other trials, chemicals were levitated off live pig skin without any apparent skin damage as noted by visual inspection by a dermatologist.[18]

CONCLUSION

We have suggested that the contamination of animal feed with a bioagent in a manner which may affect animal-derived human food products is possible, but some knowledge of microbiology and feed mill operations are required. Contamination may be possible by many types of bioagents, however, bacterial agents are most likely to be used due to the ease of purchase, relative ease of use, and general difficulty to detect in food animals so that undetected addition to human food is more likely to occur. The most important countermeasure is the rapid identification and elimination or decontamination of the adulterated feed. The luminometer and similar available instruments should be validated as an immediate screening systems to detect the bacterial contamination of feed. Test kits capable of identifying specific bioagents, such as the SMART and ALERT systems, as well as other commercially available systems, should be reviewed and validated to function in contaminated feed. The ability to rapidly develop new identification systems for unknown bioagents should be supported. Also, potential concepts of decontamination of feed should be investigated in the event of a specific feed contamination.

REFERENCES

1. CARUS, S.W. 1998. Bioterrorism and Biocrimes. Working Paper of the Center for Counterproliferation Research. National Defense University. August 1998.
2. THOMPSON, L. 1998. (Personal communication).
3. CHRISTOPHER, G.W., T.J. CIESLAK, J.A. PAVLIN & E.M. EITZEN. 1997. Biological warfare: A historical perspective. J. Amer. Med. Assoc. **278:** 412–417.
4. BEERS, A. 1998. Costco recalls beef due to *E. coli* 0157:H7 concerns. Food Chemical News July 6: 19–20.

5. Nationwide Beef Microbiological Baseline Data Collection Program: Steers and Heifers (October 1992–September 1993). 1994. Microbiology Division. U.S. Department of Agriculture.
6. SCHWARTZ, J. 1998. Change in Cattle Diet May Eliminate *E. coli* Risk. The Washington Post September 11: A3.
7. KOBETZ, P.E. & K. KOBETZ. 1994. Feed plant layout and design. *In* Feed Manufacturing Technology. **IV:** 33–48. American Feed Industry Association, Inc. Arlington, VA.
8. MUIRHEAD, S., ED. 1997. Feed marketing. Feedstuffs: Reference Issue **69:** 6–22.
9. FRANZ, D. 1999. Ann. N.Y. Acad. Sci. This volume.
10. Drug Residue Monitoring and Farm Surveillance. 1995. *In* Grade "A" Pasteurized Milk Ordinance. Appendix N. Public Health Service/Food and Drug Administration. Publication No. 229. Washington, D.C.
11. LOOMIS, L. 1998. Food Safety Monitor System with Model 3550i Microluminometer. New Horizons Diagnostics, Columbia, MD 21045.
12. SEAVER, M., J.D. EVERSOLE, J.J. HARDGROVE, W.K. CARY, JR. & D.C. ROSELLE. 1999. Size and fluorescence measurements for field detection of biological aerosols. Aerosol Sci. Tech. **30:** 174–185.
13. LOOMIS, L. 1998. General Information on all SMART Tickets for Detection of Biological Materials. New Horizons Diagnostics, Columbia, MD 21045.
14. EMANUEL, P.A., J. HASAN, R. VIN, J. DANG, L. LOOMIS & J.J. VALDES. 1998. Chromatographic immunoassays for environmental monitoring. (In press.)
15. U.S. FOOD AND DRUG ADMINISTRATION. 1998. Outline of the Center for Veterinary Medicine (FDA-CVM) data requirements for Milk Screening Tests labeled for testing milk of bulk tank/tanker trucks for drug residues. Center for Veterinary Medicine. Office for Research. Laurel, MD.
16. WALDRUP, A.L., P.W. WALDRUP & E.R. JOHNSON. 1997. Development and evaluation of a rapid test procedure to detect Salmonellae in feed ingredients and finished feeds. J. Rapid Methods Automation Microbiol. **5:** 151–167.
17. WATSON, I. 1998. Decontamination of pathogens on surface of vegetables with light energy. Personal communication. September 11, 1998.
18. ZAMANI, K., J. EZZELL, M. BONNER & M. IRANI. 1998. (Unpublished results.)

The Need for a Coordinated Response to Food Terrorism

The Wisconsin Experience

NICHOLAS J. NEHER[a]

Agricultural Resource Management Division, Wisconsin Department of Agriculture, Trade and Consumer Protection, Madison, Wisconsin 53708, USA

In late December 1996, one of the most potentially devastating events to the state's agricultural economy took place. No event in the history of the state's feed program came close to the magnitude of this incident. An unknown person or persons notified the local police chief that the feed products leaving a rendering plant had become contaminated with a pesticide and to expect large-scale animal mortality.

Shortly thereafter, the Department of Agriculture's Trade and Consumer Protection Toxic Response Team was contacted and various activities were put into place. Analytical identification tests were run on the unknown substance provided by the police chief. Samples of the rendering plant products were taken and analyzed for the identified substance and other contaminants. Federal agencies were notified of the potential problem. Strategies were developed to determine the extent and impact of the contamination and to catch who was responsible. Records were reviewed to develop a potential recall of contaminated products.

What soon became apparent was that this incident had the potential to devastate America's dairyland in terms of image and economic losses. These facts became evident in the first days of the investigation:

(1) The substance was chlordane (an organochlorine pesticide). Chlordane is environmentally stable, accumulates in the fat of animals, and is considered a food adulterant at very low levels (0.3 ppm in animal fat).

(2) Liquid fat at the plant had been contaminated by mid-December with shipments going to thousands of customers in Wisconsin, Minnesota, Michigan, and Illinois.

(3) The major recipient of these shipments was the largest feed manufacturer in the state who has among its customers more than 250 of the largest dairy farmers in the state. Ultimately it was determined that contaminated product went to approximately 4,000 farmers in a four-state area.

(4) Milk from these farms had been shipped to numerous dairy plants, which further distributed the products to other plants throughout Wisconsin and Northern Illinois to be turned into other dairy products such as cheese, butter, and ice cream.

---

[a]Address correspondence to: Nicholas J. Neher, Administrator, Agricultural Resource Management Division, Wisconsin Department of Agriculture, Trade and Consumer Protection, 2811 Agriculture Drive, P.O. Box 8911, Madison, Wisconsin 53708; Telephone: 608-224-4567; Fax: 608-224-4656;
e-mail: nehernj@wheel.datcp.state.wi.us

TABLE 1. Federal, state, and local agencies involved in investigations

City police
County sheriff
State crime laboratory
Wisconsin Department of Agriculture, Trade and Consumer Protection
  Environmental investigators
  Dairy inspectors
  State veterinarian (Animal Health Laboratory)
  Bureau of Laboratory Services
Wisconsin Department of Natural Resources
Wisconsin Department of Health
Food and Drug Administration (FDA) and their Office of Criminal Investigation
    (This office establishes a link to the FBI, Secret Service, and CIA)
United States Department of Agriculture (USDA)
United States Environmental Protection Agency (USEPA)
Departments of Agriculture in Minnesota, Michigan, Illinois, Iowa, and Indiana
Departments of Natural Resources in Minnesota and Michigan

Many agencies became involved in the incident as shown in TABLE 1. The cooperation of the feed industry was impressive. Within two days of determining that the feed was potentially contaminated with unsafe levels, all major customers were notified and the feed was replaced.

Herd samples were taken from the dairy farms with the greatest likelihood of chlordane contamination. Every sample was either negative or well below the unsafe level. Later, a few animals were slaughtered and they were found to have chlordane levels below the contamination threshold.

The recall of animal feed and liquid fat resulted in the disposal of approximately 4,000 tons of feed and 500,000 pounds of fat with an estimated value of nearly $4 million dollars.

Is this terrorism or is it an act of vandalism or revenge? A comparison to another major incident in 1981 also in Wisconsin clarifies the difference between terrorism and vandalism.

In the 1981 incident, 135 beef cattle were poisoned after an individual climbed to the top of a farm silo and placed a full bag of an organophosphate corn root worm insecticide in the silage. The silage was augured into feed bunks. The animals consumed the silage in one feeding. Most died with 24 hours and ultimately only four survived. No one was ever apprehended and it was felt that someone may have had a grudge with the farmer.

Terrorism on the other hand strives to evoke a response of fear and panic. Clearly in this case, the individual is attempting to ruin the company and, at the same time, create a high level of anxiety for the firm's customers.

In fact, the story continues. In May, 1997, a customer of the company received a letter stating that poison again had been placed in the firm's products. This time the type of poison was not identified, but the method of contamination was.

The same procedures used in the first occurrence were used again. In this instance, the products made from restaurant grease primarily went to the poultry industry instead of dairies. This included one of the country's largest turkey producers. The possible locations where the poison could be introduced numbered over 500. Once again, the information contained in the letter was accurate. However, the company had put additional safety procedures in place that included the testing for more contaminants and retaining the product longer before shipping to customers. This time a fungicide called folpet was used. The contaminated incoming material was destroyed and no feed ingredient products were contaminated in this situation.

Unfortunately, the terrorist stated that a bigger incident was scheduled to take place later in the year. Since then, additional communications have been received by the firm or its customers contending that additional problems will occur. Each of the communications comes from a different location.

A $50,000 reward has been offered by the firm. At this time, the guilty party or parties are still at large.

What have we learned? (1) Threats must be taken seriously and acted upon quickly. (2) A structure must be in place to coordinate and handle incidents of this size and scope. (3) If criminal action is suspected, there must be adequate resources and expertise to deal with the incident on two separate tracks and yet share necessary facts and evidence. One track determines the degree and extent of the contamination and protects public safety by removing the products from the marketplace. The team must be able to inspect records, sample and retain products, gather physical evidence, conduct analytical tests, collect statements, and, if necessary, dispose of contaminated products. The other track determines who caused the problem. This includes the ability to gather physical evidence, such as communications and fingerprints; conduct forensic tests, such as DNA testing; conduct interrogations and lie detector tests; possess arrest powers; secure search warrants; and work with foreign governments. (4) Feed ingredients are great vectors to cause widespread contamination and turmoil. (If the dairy industry is a key component of your agricultural industry, your economy is extremely vulnerable). (5) Current company protocols are inadequate to prevent widespread contamination if someone wants to terrorize the industry. Firms test for things they would reasonably expect to find. The costs of checking for the absurd would price their finished products out of the marketplace. (6) It could have been worse. The person could have used a different substance or a more concentrated form of the contaminant. Wisconsin could have been crippled in a massive domino effect.

The Threat of Bioterrorism to U.S. Agriculture

MICHAEL V. DUNN[a]

Marketing and Regulatory Programs, U.S. Department of Agriculture, Washington, D.C., USA

Bioterrorism, or the intentional and hostile introduction of harmful organisms and/or disease pathogens, poses a significant threat to the United States. Specifically, it threatens our country's agricultural industries, human health and safety, our economy, and ultimately our national security. In response to the threat of bioterrorism, the Department of Agriculture (USDA) is involved in federal efforts to bolster our national defenses to protect both U.S. agriculture and human safety from bioterrorism.

U.S. AGRICULTURE: STRONG BUT VULNERABLE

The agricultural industry in the United States is arguably the healthiest and most productive in the world. Because our nation benefits from an affordable and abundant food supply, the people of the United States are among the world's most prosperous. The ability of U.S. agricultural industries to provide our citizens with a constant supply of wholesome, inexpensive food is a significant factor contributing to our country's prosperity, our citizens' well being, and our national survival.

Despite the tremendous importance of maintaining the integrity of our agricultural resources, Americans have long taken for granted the blessings of a seemingly never-ending food supply. Unfortunately, we live in a world in which we can no longer afford to be so complacent—especially about something so essential to our national security. Since 1995, three incidents have occurred that provide a chilling illustration of the fact that bioterrorism poses an immediate threat: the release of nerve gas in a Tokyo subway by a religious cult and the subsequent discovery of a sizable biological weapons arsenal amassed by the group; the revelation and continued existence of an extensive biological weapons infrastructure in Iraq; and the recent discovery that the former Soviet Union's supposedly defunct bioweapons program was more elaborate and heavily funded than anyone ever believed.[1]

All of these discoveries serve as warnings that, now more than ever, our governmental infrastructures must be prepared to respond to the intentional introduction of potentially deadly biological agents. Certainly, internal government reports have highlighted the fact that throughout the United States all of us, including agriculturalists, are vulnerable to the dangers of bioterrorism.

---

[a]Address correspondence to: Michael V. Dunn, Marketing and Regulatory Programs, U.S. Department of Agriculture, Room 228-W (Stop 0109), Washington, D.C. 20250-0109; Telephone:. 202-720-4256; Fax: 202-720-5775.
e-mail: Michael.dunn@usda.gov

BIOTERRORISM: A SIGNIFICANT THREAT TO U.S. AGRICULTURE

In considering the threat bioterrorism poses, most Americans envision a direct attack on humans as the most likely scenario, and indeed the most terrifying. Just consider the worldwide response to the subway incident in Japan. But, while biological attacks on agriculture may seem to be less direct, they can be just as insidious and every bit as deadly.

The tactic of sabotaging agricultural production is not new. In World War I, the German army intentionally introduced anthrax and glanders into Romanian sheep, French calvary horses, and Argentinean livestock destined for use by Allied forces.[2] In today's world, microbiologic agents for bioterrorism are readily available. The cost and technical skill required to collect, produce, and deliver these biological agents are modest, at best. The consequences of such an introduction could, indeed would, be catastrophic.

Consider just the economic consequences of the intentional introduction of an agricultural pest or disease. Last year, U.S. agricultural industries were valued at $224 billion and generated over $1 trillion in economic activity. U.S. food producers exported nearly $60 billion worth of commodities into the international marketplace, providing millions around the world with wholesome, affordable food. Clearly, threatening this economic sector would wreak havoc on the U.S. economy as a whole.

Now consider how easily a terrorist could introduce a biological agent into a staple crop like soybeans and essentially cripple its production. The United States produced about 50% of the world's soybean crop last year. In fact, U.S. soybean production was valued at nearly $16 billion. Soybeans and soybean oil are basic components of countless food items. Soybeans are also used in livestock and poultry feeds. Because soybeans are such an integral part of the U.S. food supply, attacking soybean production would have disastrous results, such as higher food prices, trade disruption, and shifts in world livestock production.

A terrorist targeting the production of soybeans in the United States could choose from countless pathogens. One devastating exotic disease—soybean rust, for example, which occurs in Southeast Asia—could be easily introduced into the United States by placing spores into seed supplies or directly into fields of production. This particular disease would be extremely difficult to eradicate as windborne spores can travel hundreds of miles in a short period of time, thus infecting vast areas. We estimate that the introduction of soybean rust into the United States could lead to crop losses of up to 50% and could cost $8 billion to producers, processors, and consumers.

While this is a sobering scenario, the potential consequences of an animal disease introduction are even worse. Aside from the economic consequences resulting from lost productivity and lost export markets, several deadly animal diseases cause illness in humans as well. Anthrax, Rift Valley fever, salmonella, and ebola are just some of the diseases that are swift, lethal, and easily introduced by a terrorist.

Animal disease pathogens are readily obtainable from sick or dead animals throughout the world. Many universities, laboratories, and commercial companies maintain collections of them. With frightening ease, a terrorist could inject a pathogen into livestock to amplify it, draw blood from the animal, and produce a deadly serum.

Then, the terrorist could expose livestock to the disease by applying a preparation containing the pathogen through an aerosol spray. Depending on the objective, the individual could use aerial application through use of a crop duster, spraypainting equipment obtainable from any hardware store, or something as small as a perfume atomizer. One could also place a disease agent into a ventilation system in an animal enclosure.

With this in mind, let us turn to the logistics of U.S. animal production. Typically, a livestock feedlot can contain from 50,000 to 800,000 animals at one time. Poultry-raising operations can house up to a million birds. These animals live in close quarters and an infective agent would spread rapidly among them. With an infective agent causing a disease such as Rift Valley fever or avian influenza, symptoms would probably begin to appear within two to five days.

A SCENARIO OF DESTRUCTION

For the purposes of illustrating the possible ramifications of a bioterrorist attack on U.S. agricultural resources, consider a scenario in which a terrorist introduces a fast-acting virus into a population of 800,000 cattle on a feedlot. In two to three days, the span of time it would take for a significant number of animals to appear ill, as many as 10,000 head of cattle per day would have been sent from the feedlot to slaughter facilities. These animals would have mingled with other livestock and would have been handled by hundreds of humans in slaughter and processing plants. Meat from these animals would enter the marketplace, where it would be purchased and eaten by consumers.

At the same time, it is possible that anywhere from several hundred to several thousand cattle from the same feedlot would have been transported to other livestock production facilities. Again, these animals would mingle with other livestock and would come into contact with humans, infecting them with disease.

At this point, three days after the introduction, we might not recognize that a bioterrorist attack had occurred. In fact, the signs of a biological attack would come to light over time, as animals were perhaps belatedly diagnosed in isolated facilities. Rather, we would first consider other possible sources of disease, such as a sick animal or contaminated water or feed.

By the seventh day, however, nearly all of the animals would have become ill, resulting in up to hundreds of thousands of animal deaths. We estimate that tens of thousands of people would also have become ill, some lethally, as a result of handling or eating the meat. Our infrastructure would be faced with a crisis situation, and Animal and Plant Health Inspection Service (APHIS) of the USDA would be faced with the responsibility of disposing of the exposed animals in a sanitary and humane manner.

USDA: VIGILANT AND READY

All government agencies, whatever their mission, must be in a constant state of readiness to respond to the intentional introduction of biological agents for terrorist purposes. Otherwise we run the risk that our agricultural and human health response systems would be overwhelmed.

At the USDA, we have been working closely with our colleagues in other agencies government-wide to identify our vulnerabilities and overcome them. Officials from the USDA's APHIS, Food Safety and Inspection Service, and Agricultural Research Service are working with experts from the Department of Defense, Department of Justice, the Federal Bureau of Investigation, the Department of Health and Human Service's Centers for Disease Control and Prevention and Food and Drug Administration, as well as the Environmental Protection Agency.

Among the the USDA agencies, APHIS has played a traditional role in ensuring that agriculture in America remains healthy and strong. APHIS vigilantly enforces pest and disease exclusion and eradication programs. In the past, with the support of other federal, state, and private cooperators, these programs have eliminated, or brought under control, serious agricultural threats within the United States. Examples are foot-and-mouth disease and screwworm, diseases that affect livestock, and Mediterranean fruitfly, a pest that causes massive losses to the production of fresh produce. Just as APHIS stands ready to take a leadership role in responding to the introduction of naturally occurring agricultural pests and diseases, the USDA is also prepared to take action in the event that a destructive organism or disease pathogen is intentionally introduced into the United States as a hostile act.

While the USDA's expertise may be in the area of plant and animal health, it is important to note that our efforts are not exclusively focused on addressing threats to agricultural production. Accordingly, we are cooperating with other federal agencies to address other nonagricultural concerns. In this endeavor, we are greatly concerned with the threats that directly affect human health and safety. For this reason, we are taking stock of the resources we have in the animal health community to help protect human health in a crisis situation.

Returning to the scenario outlined before, if thousands of people were to fall ill in a short period of time, especially in a rural area, medical resources and personnel would soon be exhausted. When paramedics or other emergency medical technicians became in desperately short supply, animal health specialists could assist them by offering their help in stabilizing sick patients. Or, while most rural medical facilities have few respirators, many veterinary clinics and hospitals have equipment that is similar enough that with little modification it would be used to save human patients.

While the USDA is actively working with our colleagues in other agencies to devise innovative response strategies for bioterrorism, it must be emphasized that APHIS continues to develop its existing structures to keep exotic agricultural pests and diseases out of the country. These structures include rapid-response teams, extensive monitoring, wide-area monitoring, field scouts, and locally and globally networked information systems.

For example, under the Cooperative Agriculture Pest Survey program and the National Animal Health Monitoring System, APHIS collects information from state agents, individual producers, and industry groups on the occurrence and distribution of agricultural pests and diseases. This information is entered into a nationwide database that provides information on exotic pest detection and management to state and federal agencies, agricultural producers, and industry associations.

APHIS' rapid-response strategy also allows for teams of USDA specialists trained in the coordination and execution of activities associated with an outbreak to report to the site within 24 hours—regardless of whether the pest or disease of concern affects animals or plants. In such situations, APHIS teams (1) conduct activities such as testing of animals or crops and field surveys and (2) provide operational support for cooperative activities with state agencies, such as the imposition of quarantines, herd depopulation, and the application of pesticides. APHIS has entered into memoranda of understanding with every state and with other USDA agencies, as well as the Department of Defense, to cooperate in agricultural pest or disease emergency situations.

APHIS also relies heavily on its ability to prevent the entry of foreign pests by conducting agricultural quarantine inspections at U.S. ports-of-entry. In 1996, APHIS officials conducted inspections at 172 land and seaports. At airports, APHIS personnel inspected 66 million airline passengers and crew members, confiscated almost 2 million potentially damaging plant and animal products that arrived with these incoming travelers, and identified over 62,000 reportable pests.

CONCLUSION

Bioterrorism poses a significant threat to our nation's agricultural and economic viability, human health and safety, and national security. Accordingly, the United States must be prepared to respond to the intentional introduction of potentially deadly biological agents. In response to this threat, USDA is working closely with other federal agencies to monitor, identify, and safeguard areas vulnerable to bioterrorism. Furthermore, the USDA will continue to work with these agencies to plan and coordinate emergency response activities in preparation for possible bioterrorist attacks.

REFERENCES

1. PETERS, K.M. 1997. Deadly Strike: The Pentagon takes aim against chemical and biological terrorism. Government Executive Magazine **29:** 22–27.
2. U.S. DEPARTMENT OF AGRICULTURE. 1998. The threat of biological terrorism to U.S. agriculture. Animal and Plant Health Inspection Service. USDA.

Where Have All My Pumpkins Gone?
The Vulnerability of Insect Pollinators[a]

BARRY H. THOMPSON[b]

Center for Medical and Molecular Genetics, Armed Forces Institute of Pathology, Rockville, Maryland 20850-3125, USA

IMPENDING DEARTH

An urgent plea to members of a local beekeepers' association from a farmer in search of colonies for his pumpkin crop gives focus to the precarious nature of insect pollination in our country today and indicates a largely unrecognized target for bioterrorism aimed at American agriculture. "The county agent says that I won't have any pumpkins," the farmer said, "unless I get some bees for pollination. I need about ten hives tonight." None of the beekeepers contacted had any colonies available to meet his requirements.

POLLINATION

Pollination involves the transfer of pollen from the (male) anthers of a flower to the (female) pistil of the same or another flower for the production of fruit and/or seed. Plant species may be self-fertile (produce fruit or seed with their own pollen) or self-infertile (requiring cross-pollination with pollen from other plants of the same species). Some species that are self-fertile may produce increased yields or fruit or seeds of better quality if cross-pollinated. Some self-fertile species undergo auto-pollination; flowers are fertilized with pollen from their own anthers. Others have flowers that are so constructed that pollen must be physically transferred from anthers to the pistil. Some species that are self-fertile have anthers and pistils that mature at differing times, thus requiring transfer of pollen from flower to flower. Plants of these latter types and those plants that are self-infertile require a means of transfer for the pollen from one flower to another or between flowers on separate plants. Most agriculturally important grasses (wheat, barley, rice, rye, maize, and sorghum) are wind pollinated. Plant species with conspicuous flowers usually are serviced by insects or animals. For many of the 250,000 or so identified flowering plants, reproduction is dependent upon bees or other pollinators, such as bats, beetles, birds, butterflies, and moths.[1]

---

[a]The opinions expressed herein are those of the author and are not to be interpreted as representing those of the Armed Forces Institute of Pathology, the United States Army or the Department of Defense.

[b]Address correspondence to: Barry H. Thompson, Colonel, USAF, MC, Center for Medical and Molecular Genetics, 1413 Research Boulevard, Rockville, Maryland 20850-3125; Telephone: 301-319-0200 and Fax: 301-295-9507.

BEES AS POLLINATORS

The relationship between flowering plants and bees likely dates back 50–100 million years. Bees fed on the nectar and pollen of the plants and provided pollination in return. The pattern of mutual benefit continues today. Members of the Family *Apidae* (bees) possess physical and behavioral attributes needed for pollination of plants: food exchange among individuals and between adults and young, specialized mouth parts and tongues for the obtaining of nectar, a honey stomach for transport, bodies with branched hairs to which pollen adheres, and storage of pollen and/or nectar/honey in the nest.[2] Honeybees (Subfamily *Apinae*), bumblebees (*Bombini*), and solitary bees (*Euglossoni* and *Meliponiae*) are among the most effective of insect pollinators.[3] The highly eusocial European or western honeybee lives in colonies of thousands (often 50,000–75,000 bees), each headed by a single fertile female (the queen) and survives from year to year by cluster formation (in response to adverse temperatures or weather) and consumption of stored foodstuffs. Bumblebee colonies, depending upon the species, reach maximum populations in the tens or hundreds. The colonies perish each winter, save for the mated females that hibernate and emerge in the next year to establish new colonies. Among the solitary bees, *Megachile spp.* (alfalfa leafcutter bees), *Nomia spp.* (alkali bees), and *Osmia spp.* (orchard mason bees) have been the most studied with regard to commercial pollination. Mated females construct individual nests, in soil or plant stems, in which eggs are laid and the larvae provisioned by the single parent. Development of the young and subsequent emergence as adults takes place without further care by the adult. Female solitary bees each lay a small number of eggs. Some species produce multiple generations in the same season.

Honeybees have several advantages as pollinators. Colonies contain large numbers of foraging workers. The foragers may travel in a radius of up to two to three miles from the hive, thereby covering an area of thousands of acres. On a given foraging trip, a worker bee generally restricts her visits to a single type of flower; pollen transfer thus (desirably) occurs among flowers of the same variety of plant, apple pollen to apple pistil, for example. This pattern of bees has been termed "fidelity" and differs from the "casual" foraging of other insects that visit differing species of plants while fulfilling daily food requirements. Managed honeybee colonies can be moved to locations at which pollination is required; foraging honeybees can be made available during the often narrow "windows" critical for effective setting of fruit or seed.

Bumblebees and solitary bees have specialized applications in American agriculture. For some crops, they may be more efficient and better pollinators. Bumblebees, for example, can be confined to greenhouses wherein the blossoms of commercial tomatoes receive their undivided attention. The leafcutter bee is the most effective pollinator of alfalfa in farming areas in the Northwest where blossom response to climate places the honeybee at a disadvantage. Resident populations of bumblebees and solitary bees can be supplemented by those that are commercially raised for this purpose.

The complex interaction between bees and the plants that they pollinate has been the subject of intense scientific research in the last half-century. The understanding derived from these studies has been used in attempts to tailor insect presence and behavior to the desired outcome for the target crop.

TABLE 1. Plants dependent on honeybees for pollination

Alfalfa	Dewberry	Persimmons
Almonds	Eggplants	Plums and prunes
Apples	Garlic	Pumpkins and squash
Avocado	Gooseberries	Radishes
Blackberries	Herbs (spices)	Rape (oilseed)
Blueberries	Huckleberries	Raspberries
Buckwheat	Lavender	Sunflowers
Cherries	Macadamia nuts	Sweet clovers
Chinese gooseberries	Mangoes	Tea
(Kiwi)	Muskmelons	Turnips
Citrus (minor)	Cantaloupes	Vegetable seeds
Clover (Alsike, crimson, red, and white)	Casaba	Asparagus
Cole crops	Honeydew	Caraway
Broccoli	Persian melon	Carrots
Brussel sprouts	Mustard	Celery
Cabbage	Nutmeg	Chicory
Cauliflower	Onion	Chives
Kale	Parsley	Vetch
Cranberries	Parsnips	Watermelons
Cucumbers	Peaches and nectarines	
Currants	Pears	

BEEKEEPING ECONOMICS IN THE UNITED STATES

Honeybees are today the primary pollinators for a majority of insect-pollinated plants in the United States. Some 90 crops are dependent, to some degree, on bee pollination. It has been estimated that honeybees are responsible for almost $10 billion worth of fruit and seed annually. Honey production in the United States is approximately 200 million pounds annually, valued at $125 million. Almost 4 million pounds of beeswax worth $7 million are harvested annually. Thus, the value of honeybees as pollinators exceeds by perhaps 150 times the value of the honey and beeswax produced. Honeybees are said to contribute to the American diet fully one-third of all foods eaten. A wide variety of important plants, including those listed in TABLE 1, depend upon honeybees to meet their pollination needs. Certain commercial crops (listed in TABLE 2) benefit from honeybee pollination but do not require it.

Beekeeping in the United States today is a product of the technological advances of this century. The establishment of the interstate highway system and the capability of transporting hundreds and even thousands of beehives by long-haul trucks have made possible the development of commercial pollination services on a nationwide basis. For the 1998 almond crop in California, it is estimated that 500,000 to 600,000 colonies of honeybees were employed, most moving into the state for this express purpose. Colonies of bees that overwintered in the South may be transported to Virginia for pollination of apples or to the Dakotas for sunflowers. There is an extensive enterprise in the production of queens and package bees for requeening of existing colonies and for the establishment of new colonies. Queens may be shipped

TABLE 2. Commercial crops that benefit from honeybee pollination

Apricots	Coconuts	Okra
Broad beans	Coffee	Papaya
Chestnuts	Cotton	Peppers
Citrus	Cowpeas	Safflower
Grapefruit	Flax	Strawberries
Lemons	Lespedeza	Tomatoes
Oranges	Lima beans	

After Sammataro & Avitabile.[3]

from Hawaii to New York. The products of the hive (honey, wax, pollen, and propolis) enter commerce nationwide and abroad.

ORIGINS OF AMERICAN BEEKEEPING

Honeybees are not native to the western hemisphere. Tradition holds that European honeybees (probably the German black bee, *Apis mellifera mellifera* Linnaeus) were brought first to America by early colonists from England around 1622.[4] By the mid-1800s, honeybees were established throughout the temperate regions of North America; and feral colonies were as plentiful as, if not more so, than those possessed by beekeepers. Historically, honeybees had been kept in straw or wicker skeps, pottery pots, and tubes, and more recently in wooden boxes. Combs usually were built on crossed sticks in the interior and were adherent to the walls of the container. The hive contents were not amenable to manipulation, and harvesting of the honey was both difficult and destructive to the colony. Numerous attempts to improve the hive in which honeybees were kept culminated in the development in America of the movable-frame or Langstroth hive in 1851. This enabled beekeepers to manipulate the contents of the hive to control brood rearing, to minimize swarming (and thereby increase the honey crop), and to remove the surplus honey more easily for human consumption. The modular nature of the movable-frame hive also permitted the sizing of the hive for transport, an aspect critical to any significant movement of beehives for pollination. About the time that significant change came to the beehive, there was a groundswell of enthusiasm for the improvement of the honeybee itself. Breeding stocks representing the twenty-four races of bees, e.g., *A. m. ligustica* Spinola (the Italian bee), *A. m. carnica* Pollmann (the Carniolan), *A. m. caucasica* (the Caucasian), were imported as alternatives to the German bee and to develop hybrid strains. These hybrid strains possessed more desirable characteristics, such as increased gathering of honey and pollen, decreased propensity for swarming, improved winter hardiness and gentleness during manipulation of the colony. Italians, and the genetic crosses derived therefrom, remain the most widely distributed bees in the U.S. today. Attempts to produce a more productive honeybee, particularly one that is less susceptible to disease and parasitic organisms have continued to the present.

IMPACTS ON BEEKEEPING TODAY

There are presently an estimated 120,000 beekeepers in the United States, managing two and a half to three million hives of honeybees.[5] The vast majority of these beekeepers are hobbyists or backyard beekeepers, with only a small number of hives apiece. Several thousand others are "sideliners," each maintaining perhaps up to several hundred hives. Sideliners may engage in pollination of selected crops on a local basis; but for most the primary purpose of keeping bees is more likely the production and marketing of a honey crop than pollination. Several hundred individuals are commercial beekeepers, for whom pollination represents the primary source, or a significant portion, of their income.

The urbanization of the United States has reduced the numbers of hobbyist beekeepers in many areas. Apartments and townhouses do not lend themselves readily to beekeeping. The decreased lot size of modern single-family housing and increased proximity to one's neighbors exacerbates the situation. Municipal ordinances restricting beekeeping and the litigious nature of the American society further dissuade the part-time beekeeper. Of little help to beekeeping has been the public and official over-concern about the "killer bee" as the Africanized honeybee makes its way to this country from the south.

Importation of honey, in particular from Mexico and China (where the government is said to heavily subsidize production), depressed the price of U.S. honey in the early 1990s and caused major reductions in commercial consumption. Tighter labeling laws and anti-dumping legislation, limiting the importation of under-priced foreign honeys, have partially stabilized the U.S. honey market. The ultimate response of the beekeeping industry to the situation is as yet unclear.

Naturally occurring diseases and parasites afflict insect pollinators. Honeybees are subject to bacterial diseases (American foulbrood, European foulbrood, and septicemia), fungal diseases (chalkbrood and stonebrood), viruses (sacbrood, Kashmir bee virus, and F-virus), and attack by protozoans (*Nosema spp.*), amoebae, and flagellates.[6,7] Some diseases are largely self-limited. Others have been held in check by the use of prophylactic antibiotics. The nature and extent of viral diseases are poorly understood. The effects of parasites are better understood. The recent introduction of two new honeybee parasites has had a devastating impact on beekeeping in this nation.

The use of pesticides in modern farming may be the major threat to the keeping of honeybees. Bees are highly sensitive to many of the compounds currently in use. Foragers from a hive may be eliminated by contact with pesticides on flowers they visit. Entire colonies can be eliminated by indiscriminate or poorly timed applications or by the contaminated pollen brought to the hive and consumed by its inhabitants. Sometimes, in spite of the best efforts of the farmer and the beekeeper, even transported colonies are depopulated or destroyed. Fungicides, not commonly thought of as being toxic to adult bees in the same fashion as pesticides, may have as yet undetermined adverse effects on honeybee eggs and developing larvae. Gathered pollen can be the vehicle for fungicide entry into the beehive.

Beekeeping in America is subject to significant pressure from all sides. Just as societal and economic changes are making beekeeping more difficult, the demand for honeybees as pollinators is increasing, especially as the population of feral insect pollinators declines.

Decline of Wild Insect Pollinators

Significant populations of (non-honeybee) insect pollinators were present at the time of the arrival in the New World of European settlers. A decline in native pollinators has accompanied changes in farming practices, particularly those of the past fifty years. The development of large-acreage tracts and the elimination of fences and hedgerows have reduced drastically the habitat needed by insect pollinators for nesting and rearing of offspring. Herbicide use to eliminate weeds and other plants that compete with the planted crop and the development of monoculture farming lessen the availability of forage over extended periods to support wild foragers.[1] Wide-scale application of pesticides and fungicides has become the mode of most modern farming, with major adverse effect on wild as well as managed pollinators. Wild pollinators may suffer severe effects of disease and parasitic infestation due to the absence of those treatments afforded managed colonies. In some areas of the United States, native species of pollinators have been virtually eliminated. Artificial reintroduction of native pollinators (such as *Nomia*) or substitution of imported species (*Megachile*) has been attempted, with varying success. *Megachile rotundata* has been shown to perform quite well in the pollination of alfalfa in arid regions of the western United States. It remains unlikely that necessary populations of wild pollinators can be re-established and maintained, at least for the foreseeable future. Introduced pollinators generally fail to establish themselves in sufficient numbers to represent a viable replacement population, and new stocks are required on a yearly basis to meet pollination needs.

The current employment of introduced or transported pollinators by commercial producers of alfalfa (for seed), almonds, and low-bush blueberries reflects an appreciation of the means to optimize crop production by ensuring adequate pollination and of the lack of necessary local feral populations of insects.

TWO INDICATIONS OF VULNERABILITY

The Africanized Honeybee

In the late 1950s, the African bee (*A. mellifera scutellata* Lepeletier) was imported into Brazil in an attempt to improve stocks there. The subsequent release of African or African-hybrid queens resulted in the spread of the "africanized" honeybee throughout most of South America. Aggressive, prone to swarming, and requiring special techniques and equipment to manage, this strain quickly displaced the resident Italian bees, both managed and feral. Large numbers of beekeepers abandoned their craft, and beekeeping was all but eliminated in many areas.[8] Better understanding of the behavior of the africanized bee, use of protective clothing, and employment of appropriate management practices have led to a resurgence of beekeeping in recent years. Beekeeping is not what it was, but pollination needs are being met and honey produced (perhaps in larger amounts than with Italian bees in some areas.) Within four decades, africanized bees had advanced through Central America and Mexico to reach the southwestern United States. Feral africanized colonies have been identified in Texas, Arizona, New Mexico, and California. Cold winters, parasitic mites, and perhaps interbreeding with resident European stocks now appear to be ameliorating the distribution of the bee within the continental United States. However, the ultimate range of the africanized bee and its impact on American farming are still to be determined. The almost unbelievably rapid spread of the imported honeybee and the profound alterations in beekeeping which the new hybrid bees forced are clear warnings of the threat posed by introduction of a "hostile" organism into a susceptible agricultural system.

Parasitic Mites

In the past 15 years, imported parasites have become a major threat to honeybees in North America.[6,7] The tracheal mite (*Acarapis woodi* Rennie) appeared in the United States in 1983 and was associated with massive losses of bee colonies over the next few years. Apparently, the parasite was brought to this country from Europe. It is thought to have been the causative agent in the Isle of Wight disease that threatened beekeeping in the United Kingdom in the early 1900s. Tracheal mites are microscopic and infest the respiratory passages of adult bees, penetrating the walls to feed on the bee's blood or hemolymph. The physical damage done may promote secondary viral infection. Sheer numbers of mites may impede breathing. Tracheal mite populations build slowly until the number of infested bees is such that losses (incapacitation and deaths) exceed the replacement of adult bees by hatching brood. At that point, the colony begins to "dwindle." Effects of mite infestation are greatest in the winter when honeybees are clustered (the close proximity to one another promotes transfer of mites from bee to bee.) The cluster declines in size, often to a few hundred bees and their queen. Unable to maintain proper temperatures within the cluster, the bees perish. The Varroa "mite" (actually the arachnid, *Varroa jacobsoni* Oudemans), a parasite of *Apis cerana* in southern Asia, was found in America about 1987. Varroa feeds on the hemolymph of developing larvae (often causing physical deformity in the resultant adult bee) and adult bees. Loss in a colony infested with Varroa usually occurs precipitously. Populations of mites reach maximal levels during the main brood rearing season (summer). The number of infested bees reaches a threshold at which the colony "collapses." The bees die or may abscond, leaving their hive for another location. Populations of these parasites can be reduced by use of chemical agents (menthol and fluvalinate). But the chemicals must be applied according to strict schedules to be of benefit and to avoid contamination of honey intended for human consumption. Reports of mites developing tolerance to chemicals have recently begun to be published.[9] Genetic resistance to the parasites appears to be the ultimate solution, but it has yet to be achieved in on-going breeding programs. Even with the initiation of effective treatment programs, the two parasites have caused immense losses of managed honeybee colonies. In 1995, some commercial outfits reported destruction of 50–75% of their hives. Many smaller beekeepers were wiped out. Feral colonies of honeybees have been all but eliminated by tracheal mites and Varroa. The figures vary, but the Northeastern United States may have lost 80–95% of wild stocks.[5] In the instance of tracheal and Varroa mites, the introduction of parasites into a susceptible population resulted in exceedingly high mortality in a relatively short period of time.

AN INVITING AND LARGELY UNAPPRECIATED TARGET

Managed insect pollinators, inasmuch as they are the major, if not sole, source of pollinators for many economically important crops in the United States, seem to present a largely unappreciated opportunity for a naturally occurring or intentionally introduced threat to American agriculture. Drastic reduction or elimination of the honeybee would severely damage fruit and seed production and the provision of alfalfa hay/forage, the most important food for herbivores used for meat and milk. There are many ways to attack the honeybee.

The negative impact of the africanized bee on beekeeping in South America and the devastation resulting from the introduction of two parasites into North America suggest that utilization of other naturally occurring organisms could result in similar or more far-reaching insults. Such organisms, once identified, likely could be obtained without major effort or expense. Introduction alone would probably suffice as a means of attack as exemplified by the natural spread of tracheal mites throughout North America in just a few years.

Naturally occurring bacteria or viruses or protozoans could also be employed. American foulbrood, because of its ability to form spores that are resistant to chemicals and capable of disease production after decades of inactivity, remains a threat to beekeeping. Infected bees, brood, and wax combs are best destroyed by burning. Infected wooden hives can be cleaned by scorching the interior. For reasons that are not fully understood, the occurrence of American foulbrood has declined in the United States in recent years. Nonetheless, a mutated (or genetically engineered) form with increased virulence could be a significant new means by which to damage pollination.

A variety of bee viruses have been identified. Several cause "bee paralysis" and death of infected bees. The effect of several viruses is unknown. Recent studies suggest that concomitant infestation/infection with Varroa mites and bee viruses results in the "parasitic mite syndrome," in which synergism may accentuate symptoms of viral infection. Difficult to diagnose with certainty, bee viruses may be another avenue of attack. Selection of a naturally occurring virus seems the most likely route to follow.

The losses due to pesticide toxicity invite consideration of direct poisoning of honeybees, but this would be difficult to achieve on a scale large enough to disrupt beekeeping. More insidious, less likely to be detected, and yet reasonably easy to carry out would be the utilization of micro-encapsulated formulations of pesticides. Micro-capsules approximately the size of grains of pollen have been shown to be gathered from plants by bees and carried to the hive and stored in the manner of pollen. The toxic nature of micro-encapsulated pesticides for bees, primarily to brood being fed pollen, has been documented. Dusting of crops or aerosol distribution of micro-capsules would expose large numbers of insects (and their parent colonies) to insult. It has been suggested that honeybees be used to distribute *Bacillus thuringiensis* (a bacterium active against the larval stage of certain crop pests) to plants. Honeybees from a hive treated with micro-encapsulated pesticide could be used to spread the agent to flowers, which are then visited by foragers from untreated hives, which subsequently contaminate their own hives.

Different modes of attack include sterilization of males (drones) to disrupt reproduction, enhancement of lethal sex allelism, use of "pheromones of confusion," and hormonal disrupters. Sterilization of males has been successfully employed in control of the cattle screwworm. Were male bees to be sterilized in numbers sufficient to compete successfully with non-sterile males in mating with (flying) virgin queens, fertility in the queens would be reduced. As the sperm stored in the queen bee's spermatheca are used to fertilize eggs (which then develop into worker bees or replacement queens), the presence of "blank" sperm would adversely affect the production of worker bees, the foraging population that performs pollination. The sex-determining locus in honeybees has multiple alleles, some of which are lethal in individuals that are homozygous for the allele.[10] Hypothetically at least, concentration of lethal sex alleles might be approached by artificial insemination/inbreeding. The presence of alleles in such pattern as to enhance homozygosity at the sex-determining locus would result in reduced numbers of viable offspring.

Scientists are just beginning to understand the chemical communicators or pheromones by which honeybees maintain hive integrity and accomplish tasks of the colony. The mandibular gland pheromones of the queen are critical to the recognition of the presence of the queen, to the suppression of ovarian development and egg-laying by worker bees, and to the prevention of swarming. The pheromone of the Nasanov glands of worker bees guarding the entrance identifies the individual hive. Masking these pheromones or artificially introducing pheromones at other than normal colony sites could disrupt colony function. It has also been suggested that pheromones could be applied to plants targeted for pollination, thereby increasing the chance that foraging worker bees will pollinate the flowers of the chosen species. In reverse application, misdirection of foragers (and thus failure to pollinate the chosen species) could result from the artificial presence of the pheromone on non-selected plants.

Division of labor within the hive seems to be based upon age. Some investigators[11] propose that juvenile hormone levels (which modify behavior) rise, not in response to age, but rather in response to external stimuli — that division of labor mirrors perceived need for task completion. Hormonal disrupters might be employed to adversely modify behavior of individual bees and the colony as a whole.

The more popularly envisioned means of negatively impacting American agriculture as a whole or in great part, such as thermonuclear weaponry, neutron bombing, or application of electromagnetic forces, obviously would affect the human population as well. The latter likely would be the target, rather than the agricultural base. Thus, these types of attack have been excluded from discussion here. The effective disruption (other than perhaps by a national shortage of vehicular fuels) of the commercial transportation of pollinators, which would require Herculean effort and be unlikely to succeed, also has been eliminated from consideration.

CONCLUSION

While there has not been a documented intentional attack on honeybees or other pollinators, the potential is there. The dependency of a large portion of agriculture upon these creatures is an obvious vulnerability that bears real consideration as the target for the sophisticated bioterrorist. Effective countermeasures are presently lacking.

Awareness is the primary requirement of any program to address this vulnerability. Threat assessment by experts in apiculture, agriculture, economics, and terrorism/counter-terrorism would follow. The third step would be the development and implementation of plan(s) for minimizing vulnerability, improving detection and identification of naturally occurring or intentional threat activities, and bringing countermeasures to bear on the threat.

There is a fairly large beekeeping infrastructure in place in America. All states and many counties have beekeepers associations. Many universities and state agricultural extension services have programs that involve bees and beekeeping. The American Beekeeping Federation, the American Honey Producers Association, and the Eastern and Western Apicultural Societies are national organizations, encompassing individuals from all segments of the beekeeping community. There are more academically oriented organizations such as the American Association of Professional Apiculturalists. The American Honey Board promotes honey to consumers and addresses consumer issues.

The United States Department of Agriculture and its components, the Agricultural Research Service (and its laboratories at several locations), and the Animal and Plant Health Inspection Service are the lead agents in addressing the aforementioned issues at the level of the federal government. Collaborative interaction among interested and affected parties must be encouraged. Funding appropriate to program definition, development, and implementation will be required. Supportive legislation and dedicated long-term funding must follow if the program is to be effective.

REFERENCES

1. FREE, J.B. 1993. Insect Pollination of Crops. 2nd Edit. Academic Press. London.
2. HOOPINGARNER, R.A. & G.D. WALLER. 1992. Crop Pollination. *In* The Hive and the Honey Bee. J.M. Graham, Ed. Dadant and Sons. Hamilton, IL.
3. SAMMATARO, D. & A. AVITABILE. 1998. The Beekeeper's Handbook. 3rd Edit. Cornell University Press. Ithaca, NY
4. SMITH, D.A. 1977. The First Honeybees in America. Bee World **58**(2): 56.
5. MARTZ, E. 1998. Where Have All the Bees Gone? Pennsylvania State Agriculture. Winter-Spring. http://aginfo.psu.edu/PSA/ws98/bees2.html
6. BAILEY, L. & B.V. BALL. 1991. Honey Bee Pathology. 2nd Edit. Academic Press. London.
7. SHIMANUKI, H. *et al.* 1992. Diseases and Pests of Honey Bees. *In* The Hive and the Honey Bee. J.M. Graham, Ed. Dadant and Sons. Hamilton, IL.
8. SPIVAK, M., D.J.C. FLETCHER & M.D. BREED. 1991. The "African" Honey Bee. Westview Press. Boulder, CO.
9. ELZEN, P.J. *et al.* 1998. Fluvalinate resistance in *Varroa jacobsoni* from several geographic locations. Amer. Bee J. **138**(9): 674–676.
10. WINSTON, M.L. 1987. The Biology of the Honey Bee. Harvard University Press. Cambridge, MA.
11. SEELEY, T.D. 1995. The Wisdom of the Hive. Harvard University Press. Cambridge, MA.

The Role of Pesticides in Agricultural Crop Protection

NANCY N. RAGSDALE[a]

*Agricultural Research Service, U.S. Department of Agriculture,
Beltsville, Maryland 20705-5140, USA*

INTRODUCTION

The goal of agriculture is an available, nutritious, and affordable food and fiber supply. In the United States, we enjoy the benefits of a strong agricultural production system. It is estimated that 15% of the gross domestic product is based on the production of food and fiber.[1] A century ago, agriculture was the predominant way of life; currently, less than 2% of Americans are directly involved in agricultural production. Yet we have greater agricultural abundance than ever. Technological advances and good management practices have had important roles in enormously increasing agricultural production while reducing manpower needs. In 1996 there were 2.06 million farms occupying 968 million acres.[2] Preliminary figures for 1996 indicate that farm product exports were valued at almost $60 billion or 10% of total domestic exports.[3] In 1996, U.S. consumers spent 10.9% of their disposable personal income on food[2] compared to 20.6% in 1950.[4] The per capita consumption of fruits and vegetables in 1995 was approximately 700 pounds (based on farm-weight equivalent); the per capita consumption of grain (excluding quantities used in alcoholic beverages and corn sweeteners) approached 200 pounds.[2]

Agricultural production is not natural. Since the distant past when man started producing food rather than hunting and gathering it, there has been a continuous struggle with nature. Nature maintains ecosystem stability. When a new element, such as a crop, is inserted into an ecosystem, the components of that system shift in response. Thus our agricultural production systems, artificial additions to the natural environment, are constantly challenged, and pests are a key contender. Pests have had strong impacts in the making of history. One of the most notable events was the potato famine in Ireland, resulting from a disease caused by a fungus, causing starvation and mass migration from that country. Frequently native pests are the problem, but accidentally introduced foreign pests can also wreak havoc. Some of the more infamous introductions are the Mediterranean fruit fly, the gypsy moth, kudzu, and more recently, the Asian long-horned beetle. The primary agricultural pests are usually categorized as weeds, insects, or fungi. Other pests include vertebrates (rodents, birds, etc.), nematodes, bacteria, viruses, and so on.

Adequate pest-control methodology is necessary for the United States to compete in domestic and foreign markets. Chemical, genetic, biological, and cultural controls

---

[a]Address correspondence to: Nancy N. Ragsdale, Agricultural Research Service, U.S. Department of Agriculture, Beltsville, Maryland 20705-5140; Telephone: 301-504-4509; Fax: 301-504-6231.
 e-mail: nnr@ars.usda.gov

are all used to reduce crop losses caused by agricultural pests. The best method of pest management is to use crops that are naturally resistant to the pest. If such stock is not available or if the resistant qualities are lost, biological control using non-pest species is an alternative. These biological pest control agents may poison, infect with disease, literally eat, or cause other lethal effects to the pest population. Cultural control, such as crop rotation, is an effective pest management tool. Chemical pesticides are frequently the most economically feasible method to reduce pests. Pest management is a dynamic area that is an ongoing challenge that scientists must continue to address. Ever-changing conditions caused by local climates, new crop varieties, and fluctuations in pest pressure, related to changes in pest populations, give chemical pesticides a key role in successful pest management systems.

USE OF PESTICIDES

Various chemical pest control strategies have been tried over the centuries. The ancient Greeks used sulfur as a chemical pesticide to control plant diseases caused by fungi, particularly on stored commodities.[5] In the twentieth century, and primarily during the last fifty years, we have seen the development of synthetic chemical pesticides. They now play an integral role in American agriculture and are a vital part of the U.S. economy, increasing yields and substituting for labor, machinery, and fuel. In 1995, pesticide purchases in the U.S. represented approximately one-fifth of the

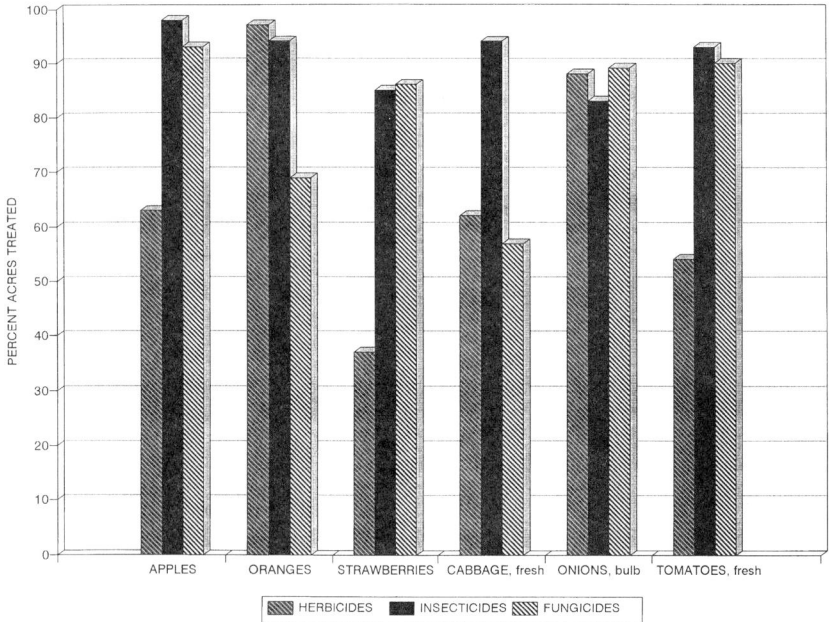

FIGURE 1. Percent acres of the respective commodities in major producing states that receive field applications of pesticides.

quantity of active ingredient sold worldwide and about one-third of the world market in dollars.[6] The United States spends about 7.5 billion dollars per year for agricultural pesticides; herbicides account for about two-thirds of that,[7] reflecting their widespread use on large-acreage, major crops such as corn and soybeans. Pesticides not only play an important role in production of major crops, such as corn, cotton, soybeans, and rice, they are also critical in production of fruits and vegetables, as illustrated in FIGURE 1. This information (FIG. 1) represents some of the data released by the U.S. Department of Agriculture (USDA) for the 1995–1996 survey data collected and analyzed by the National Agricultural Statistics Service.[3] These figures are for field application only and do not reflect seed or post harvest treatments. Currently trends point to an increasing reduction in the variety of pesticides available to control pests in agricultural production systems, particularly for the "minor" crops, such as fruit and vegetables. "Minor" use refers to U.S. crops grown on 300,000 acres or less. But the term also applies to some low use patterns that are not supported by the registrant company since sales volume does not offset registration costs.

BENEFITS OF USE

Today's farmers produce over 80% more per acre than was produced by the preceding generation and part of this increase is due to pesticides.[8] Pesticides provide benefits to producers through prevention of yield losses, improved crop quality, enhanced market opportunities, facilitation of farm work and harvest, and an improved cost/profit ratio. Consumers, in turn, benefit from a food supply that is ample, varied, safe from microbial contaminants, available throughout the year, and promotes good health. However, agriculture has not taken the steps necessary to present credible information that reflects the benefits derived from pesticide use. In order to examine these in a more definitive manner, one must examine the impacts of changing pest management tactics.[9] This examination involves such concepts as economic thresholds, injury levels, and crop loss. The data needed are focused on specific crops and pesticide(s) in question and include: (1) use data: acres grown, acres treated with pesticide(s) under review, rate applied, method and number of applications, and target pest(s) of economic importance; (2) use-associated factors: crop production cost, crop yield, crop quality, crop production price, and use of the pesticide(s) in pest resistance management and integrated pest management; (3) biological impact resulting from cancellation of specified pesticide(s): identification of alternative pesticides and/or practices and changes in yield, quality, and costs that occur when using alternative practices; and (4) economic impact analysis: based on use data, use-associated factors, and biological impact of cancelled pesticides with net effect determined by Marchallian demand-and-supply curves. The economic impact data are particularly difficult to generate and we need protocols that permit scientifically acceptable impact assessments.

REGULATION

A pesticide is defined as any substance or mixture of substances intended for preventing, destroying, repelling, or mitigating any pest. Also included in this definition is any substance or mixture of substances intended for use as a plant regulator, defoliant, or desiccant. Pesticides are biologically active compounds and, although many of the more recently developed pesticide chemicals don't kill pests outright, they interfere with metabolic processes. As a result they might pose risks to humans and other living organisms. Over a period of almost forty years, concern about pesticide use and the side effects on man and the environment have prompted safety requirements governing use. Until safety and environmental issues became public concerns in the 1960s and early 1970s, the regulatory framework was primarily motivated by the desire to prevent consumer fraud and protect legitimate manufacturers. Registration of pesticides is under the jurisdiction of the U.S. Environmental Protection Agency (EPA). Public perceptions about pesticides in general and the resulting regulatory climate have not been favorable toward maintaining current registrations or developing new ones.

EPA regulates pesticides and their use through two primary statutes: the Federal Insecticide, Fungicide, and Rodenticide Act (FIFRA) and the Federal Food, Drug, and Cosmetic Act (FFDCA). Not all the restrictions on pesticide use result from FIFRA and FFDCA regulations. Other major federal statutes relevant to pesticide use include the Endangered Species Act, the Safe Drinking Water Act, the Clean Water Act, and the Clean Air Act. To give an example of restrictions from one of these acts, the Clean Air Act carries provisions that address potential ozone layer depleters and will result in loss of the majority of uses of a major pest control chemical, methyl bromide, effective January 1, 2005. Methyl bromide has been used widely as a fumigant for controlling a wide variety of pests in soil, stored commodities, structures, and shipments that must meet agricultural quarantine regulations.[10] The loss of methyl bromide will have significant economic impacts on producers and consumers in addition to raising concerns about pests entering the U.S. with imported commodities. Currently research efforts are underway to develop new and effective alternatives.

The public's perception of risk is a large factor in considerations related to the registration status and use of pesticides. In the United States, pesticide issues seem to have become focused on food safety from the perspective of residues and their potential for adverse effects on human health. The Food Quality Protection Act (FQPA) of 1996 amended FIFRA and FFDCA, making significant changes in the process of registering and maintaining registration of a pesticide use. It repealed application of the Delaney Clause, which mandated certain approaches in determining cancer risk resulting from exposure to pesticides. In the initial registration process under FQPA, pesticides that meet qualifications as reduced-risk or minor use pesticides will be given a faster review than traditional pesticides. Once a pesticide is registered, it will be re-examined every 15 years to determine if it meets the scientific requirements for registration at that time. EPA must also assess all currently registered pesticides to determine if they meet the new standards. Benefits of pesticide use are no longer a major consideration in the process;[9] however, benefits information will remain important as a mechanism to select efficient pest management systems, in situations comparing pesticides in the risk mitigation process, and for the agricultural community to inform the public about the ramifications of losing pest management tools.

FQPA established uniform standards for setting tolerances (maximum pesticide residue allowed) in raw and processed commodities. In setting tolerances, many factors are taken into account. Based on animal feeding studies (frequently rats), safety factors are applied to account for differences in animals and humans, human variability, and now, according to some interpretations of FQPA, variability between adults and infants/children. This results in an overall 1,000-fold safety factor, and there is a great deal of discussion on whether a 100-fold safety factor would be sufficient in many cases.

EPA has used what they call the risk cup to illustrate how they will set tolerances under FQPA. This risk cup, when filled, represents the acceptable exposure to a specific pesticide. This is the amount of that pesticide to which one could be exposed every day for 70 years without detrimental effects. Before FQPA, tolerances were based on exposure through food intake and there was a separate risk cup for each pesticide. Now, in addition to exposure through food, EPA must consider exposure from water and non-dietary sources such as the home, garden, and pets. Adding up exposure levels to a specific pesticide from all these sources is called the aggregate exposure risk. In addition, EPA must also take into consideration exposure to all pesticides with a common mechanism of toxicity. In the final analysis, EPA will combine the aggregate exposure to a specific pesticide with the aggregate exposures to all other pesticides with a common mechanism of toxicity and determine if the risk cup is full. If the risk cup runs over, there are a number of options. Label changes could be made to reduce the dose, reduce the number of sprays, lengthen the time from spray to harvest, and so on. More room could be made in the risk cup by saving those uses that provide greater economic returns and eliminating minor or speciality uses.

By August 1999, FQPA required EPA to develop and put into place a screening program to gather information about the potential for pesticides and other substances to disrupt the human endocrine system. Endocrine disruptors include compounds that block or mimic the effects of hormones or act on the endocrine system in such a way they may cause reproductive or developmental problems. FQPA also requires EPA to publish and provide information to large retail grocery stores explaining the risks and benefits of pesticides, how the government protects consumers, and recommendations to consumers for reducing dietary exposure to pesticides while maintaining a healthy diet. The grocers are allowed to decide how the information will be made available to the public. There is no penalty for failure to display the information.

IMPORTANCE OF RESEARCH

Research has made dramatic advances in chemical pesticides, producing compounds that are more target specific, can be applied in smaller quantities, and are less persistent in the environment. In addition, there are more effective application and disposal methods. Research needs have been emphasized with the implementation of FQPA. Prioritization should be placed on research that can bring more scientific data into the decision-making process. This research includes mode of action, residue analysis and supporting methodology, environmental fate, metabolic fate, human exposure, and development of various models to provide more accurate estimates. Risk-mitigation research can lay the groundwork for reduced-risk pesticides and includes studies in the areas of pest biochemistry and physiology, pest-ecosystem relationships, and natural products with pest management properties. Risk mitigation can also be achieved through research that contributes to improved application delivery, disposal technology, and pest management/cropping systems.

The continuing availability of a variety of pesticides to adequately manage agricultural pests is a subject for concern. The U.S. has had a noticeable decline in public sector research and expertise in the area of pesticides over the past two decades. There are a number of cases in which national governments have curtailed funds for research on pest control chemicals.[11] Questions have been raised whether biotechnology and integrated pest management will obviate the need for synthetic chemicals to manage pests in agriculture. Concern about research in the public sector related to agricultural pest management chemicals led the U.S. Department of Agriculture (USDA) to request a study through the National Academy of Sciences, the National Research Council, Board on Agriculture on the future of pesticides in pest management for U.S. agriculture. This study will identify circumstances that might require chemicals, determine what types of chemical products would be most likely tools for the future, and recommend an appropriate role for the public sector in the research and education that underlie the development and use of agricultural pesticides. The study should be completed by December 1999.

CONCLUSIONS

Farmers are faced with the continuing problem of how to raise income and maximize production in the face of increased costs and restricted chemical pest-control materials.[12] Intelligent use and a basic understanding of pesticides are key factors to their safe use and continued availability. The public perception of risk, which has a strong influence on pesticide availability and the research that undergirds it, can be improved through better communications between the scientific community and the public. The evaluation of pesticide risk must shift to a process based on scientific evidence rather than a political process strongly driven by perceptions.[13] Scientists must play a more active role in policy decisions that involve scientific information. Such decisions are often dominated by legal expertise, economists, regulators, and policy analysts. The scientific community must assert itself in stressing the importance of continuing research in all aspects of pest control and the use of the best available scientific information in decision-making processes affecting agricultural production. Research must continue to play active roles both in devising reliable plant health management systems and in policy decisions related to implementation of these systems.

REFERENCES

1. GLICKMAN, D. 1998. Washington Post Sept. 20: A3.
2. USDA, OFFICE OF COMMUNICATIONS. 1997. Agriculture Fact Book 1997. Government Printing Office. Washington, D.C.
3. USDA, NATIONAL AGRICULTURAL STATISTICS SERVICE. 1998. Agricultural Statistics 1998. Government Printing Office. Washington, D.C.
4. CLAUSON, A. & A. MANCHESTER. 1998. USDA, Economic Research Service. Unpublished data.
5. MCCALLAN, S.E.A. 1967. History of fungicides. *In* Fungicides: An Advanced Treatise, D.D. Torgeson, Ed. **I:** 1–37, Academic Press. New York.

6. ASPELIN, A.L. 1997. Pesticide Industry Sales and Usage—1994 and 1995 Market Estimates. EPA, Office of Prevention, Pesticides and Toxic Substances. Washington, D.C.
7. ANDERSON, M. & R. MAGLEBY, EDS. 1997. Agricultural Resources and Environmental Indicators. Agricultural Handbook No. 712: 116–134. USDA, Economic Research Service. Washington, DC.
8. RAGSDALE, N.N. 1987. Agricultural pesticides: can we reduce the risks? Fertile Fields **2:** 9–11.
9. RAGSDALE, N.N. & R.E. STINNER. 1999. The role of benefits in the regulatory arena. *In* Pesticides: Minimizing Risks and Optimizing Benefits. N.N. Ragsdale & J.N. Seiber, Eds.: 156–164. American Chemical Society. Washington, D.C.
10. RAGSDALE, N.N. & W.B. WHEELER. 1995. Methyl bromide: risks, benefits and current status in pest control. Rev. Pestic. Toxicol. **3:** 21–44.
11. KLASSEN, W. 1995. World food security up to 2010 and the global pesticide situation. *In* Eighth International Congress of Pesticide Chemistry-Options 2000. N.N. Ragsdale, P.C. Kearney & J.R. Plimmer, Eds.: 1–32. American Chemical Society. Washington, DC.
12. BOSSO, C.J. 1987. Pesticides and Politics. University of Pittsburgh Press. Pittsburgh, PA.
13. FLORA, C.B. 1990. Pesticide risk: making decisions. Plant Dis. **74:** 105–108.

The Status and Role of Vaccines in the U.S. Food Animal Industry

Implications for Biological Terrorism

PETER L. NARA[a]

Research and Development, Biological Mimetics, Inc., Frederick, Maryland 21701, USA

INTRODUCTION

Domestic livestock and the food animal industry worldwide are a source of livelihood and survival for people in almost every part of the world. Animals provide food, clothing, fertilizer, fuel, shelter, companionship, transportation, and animal power to cultivate crops. The worldwide distribution of livestock includes more than a billion cattle, 767 million hogs, and 941 million sheep.[1] Interestingly, the U.S. has only 8% of the world's livestock population but produces meat, meat products, milk, and milk products equivalent to all that produced by the less developed countries. Total personal consumption expenditures for food in the U.S. is 13.1%, the lowest by far in the world. In general, the public has taken this for granted. The public also has no appreciation for what a natural or intentional introduction of a foreign animal disease (FAD) would do to the availability and cost of food from this source.

Infectious disease and severe climate changes remain major challenges to maintaining these food animal populations (TABLE 1). In the United States, food animal health and production depend on the control of indigenous diseases (IDs) and the exclusion of highly transmissible FADs.

A targeted introduction of a FAD into the U.S. would substantially threaten the U.S. agricultural economy by disrupting markets. In 1992, the value of U.S. livestock and poultry was estimated at $80 billion and approximately 20% of our gross domestic product. The United States exports hundreds of millions of metric tons of meat each year worth approximately $6.5 billion in 1990 dollars.[1] A resurgence of FADs appears to be taking place in some parts of the world, particularly in less developed regions. Some of the most ravaging livestock diseases now exist in areas where they have never been before or are recurring in places where they were once eradicated. Foot-and-mouth (FMD) disease, according to the United States Health Association, is probably the most infectious and economically devastating disease known and is still found nearly worldwide. Most of South America is affected by FMD. The disease has not been seen North America since 1954, when it was eradicated from Mexico. Based on a study some years ago, if FMD were reestablished in the United States, it would cost producers an estimated $12 billion over a 10–15-year period and increase the cost of meat and dairy products by more than 25%.[1,2]

---

[a]Address correspondence to: Peter L. Nara, D.V.M., Ph.D., Director of Research and Development, Biological Mimetics, Inc., 431 Aviation Way, Frederick, Maryland 21701; Telephone: 301-620-7691; Fax: 301-694-7223.
   e-mail: nara@sri.org

TABLE 1. Major threats to livestock industry

Indigenous infectious diseases
Foreign animal diseases
Increasing antibiotic resistance
Climate/pests as primary and secondary effects on animals and their foodstuffs
Nutrition fads and changes in consumers attitude
    What is "healthy"
    Chemical/hormone residues in meat and milk
Intensive production/husbandry practices, overcrowding/stress-related disease
Genetics of feed conversion versus resistance to disease versus excessive physiologic traits

TABLE 2. Circumstances favoring agricultural bioterrorism

Geographic consolidation and dense population animal husbandry practices of the three major food animal species (cattle, poultry, and swine)
Large number of microorganisms and toxins to select from (~120 indigenous diseases and 56 FADs)
The READO system is designed to address small, limited focal-point source outbreaks of natural/unintentional introduction. In the case of zoonotic organisms there is no full integration with the Centers for Disease Control, the Department of Defense, or the National Institutes of Health

As changes in the U.S. livestock and poultry industry are increasing the vulnerability to the introduction and spread of an FAD increases (TABLE 2).[2–4] Intensive farming and the advent of much larger production units, which take advantage of the financial gains from economy of scale and production efficiencies, have created a more favorable environment for FADs. One striking change is that farms are decreasing in number yet increasing in size. In 1977, the United States had almost 2.54 million farms; in 1990, the number of farms was less than 2.1 million.[1] Today, it is not unusual to see a flock of more than a million chickens, a feedlot of more than 100,000 animals, and a dairy herd of several thousand cows. Livestock industries have also evolved into smaller geographic patterns.[2] Animals are moved more frequently to maximize economic and seasonal gains. For example, dairy calves are moved from the Midwest to the Southeast, and back again. Feeder cattle may be reared in the South but grazed, backgrounded, or fed in several locations across the country. Another change is that these animals have been genetically selected and bred for their high rate of weight gain and feed conversion, not for disease resistance. In some cases, animals are now more uniformly susceptible to diseases.

GENERAL CONTROL MEASURES

IDs and FADs constitute a long list of pathogenic microbial agents ranging from viruses to bacteria to parasites to fungi and transmissible "prion" diseases. As opposed to an FAD, which is not present in the United States, a reportable disease is any condition found in target animal species for which state and/or federal authorities must be notified. A reportable disease may be an FAD or ID that is on an eradi-

TABLE 3. High priority bioagents for use in food animal terrorism

Poultry: Avian influenza and velogenic viscerotropic Newcastle disease (exotic Newcastle disease)
Cattle: Foot-and-mouth virus and rinderpest (cattle plague)
Swine: African swine fever and hog cholera

cation list or surveillance list. Currently 56 FADs are classified and prioritized in the United States. An emergency program have been devised for each FAD.[1] Other FADs and IDs diseases exist that may require future consideration and study (e.g., influenza). The emergency programs of the Regional Emergency Animal Disease Eradication Organization Structure (READO) under the coordination of the National Veterinary Services Laboratory of the Animal and Plant Health Inspection Service (APHIS) are designed to address one disease or a few outbreaks of a disease at one time, based on previous experiences. But it should be noted that these programs are designed for the unintentional introduction of most FADs. To date there has been no known intentional introduction of animal disease to our country's livestock and no specific programs have been designed to respond to an intentional introduction in the United States.

A well-planned biological attack could be part of a multiphase initiative aimed at overwhelming and crippling our public-health programs and livestock industry. If subtly and properly done, a biological attack on humans and animals would result in a major erosion of confidence in both human health care and food sources.[2] The intentional introduction of three or four highly contagious and fatal animal diseases would overwhelm our current system. If any one of these animal diseases had a zoonotic potential a significantly greater impact and disorganization of social activities would occur (TABLE 3).

Currently, vaccination serves as a more significant preventative barrier to IDs than FADs. Eradication through slaughter of both infected and exposed animals remains the method of choice for FADs in contained outbreaks. If an FAD becomes widely disseminated and the standard depopulation policy is not practical, vaccination, if available, may be employed to reduce losses and limit the spread of disease.

VETERINARY VACCINE FACTS

In contrast to human vaccine considerations, veterinary vaccines must meet the requirements of being relatively cheap and easy to administer in the field. The cost associated with many veterinary vaccines must be in the range of $0.10–0.50 in the food animal industry and in the range of a dollar per dose in the pet animal industry. Because there have been problems developing effective vaccines for a number of viral, bacterial, and parasitic pathogens, costly (but effective) surveillance, detection, reporting, quarantine, and depopulation (SDRQD) infrastructures have been created. The SDRQD type of disease prevention structure must not interfere with the research and development of new and effective measures (i.e., vaccines) that promote disease reduction and protection of the national herd.

Veterinary vaccines are used to fight a wide array of viral, bacterial, parasitic, and mycotic microorganisms and their related toxins. The estimated annual production

TABLE 4. Characteristics of Class II pathogens

Natural history of disease fails to induce spontaneous recovery (chronic active infections) or long-lasting immunity

Natural history of disease demonstrates pathogen's capacity to re-colonize host at varying periodicity

Pathogen exhibits antigenic variability (genetic instability)

Early innate or adaptive immune responses are robust, misdirected, cross-reactive, non-protective, and/or type-restricted

Prime/boost with unmodified antigen elicits non-protective, disease enhancing, suppressing, or type-specific responses with specific memory

TABLE 5. Characteristics of Class I pathogens

Natural history of disease in a given population exhibits spontaneous recovery and durable immunity

No, or limited, antigenic variation occurs (genetically stable)

Induction of a normal primary/secondary immune response leading to effective cell-mediated and humoral immunity with memory

of these vaccines/biologics registered with the USDA's licensed Veterinary Biologics Group in 1997[5] for all species ranged from 63 billion doses total for vaccines and viruses, 1 billion doses for bacterins, 164 million doses of combination bacterins/toxoids, and 25 million doses of toxoids. It should be noted that these numbers represent doses for both food and companion animals in the United States only and were not all necessarily used in the vaccination of these species.

There appears to be two situations in which the U.S. veterinary vaccine preparedness is compromised. First, some of the pathogens reviewed in this paper, and listed in TABLE 3, constitute a group of microbial pathogens known as Class II type (TABLE 4). These pathogens (avian influenza and foot-and-mouth viruses) generally do not have effective vaccines because they mutate and change their surface from year to year. Class I pathogens (TABLE 5), in contrast, are significantly more stable genetically and have effective vaccines (Newcastle disease virus, rinderpest, and hog cholera). African swine fever appears to be genetically a Class I pathogen, but the immune response it induces is very meager.

POULTRY

Influenza

Avian influenza, otherwise known as fowl plaque, is a single-stranded RNA virus in the Family Orthomyxoviridae and affects both domestic and wild birds.[1] The virus is an example of a Class II viral agent, has four major genera, and can cause 90–100% mortality. Viral strains exhibit a wide range of pathogenic outcomes from subclinical to mild respiratory syndromes to loss of egg production to high mortality. The virus is transmitted to domestic birds in nature by wild birds (particularly waterfowl). Fecal material (as little as 1 gram) from these "carriers" has been found to contain as many as 10 million viruses. Recently, a strain of avian influenza A

(H5N1) was discovered in birds in Hong Kong. Within one month, six people were dead from the influenza. A total of 18 cases have been reported to date. Recent research suggests that these "avian" strains may acquire the potential for infecting humans through genetic reassortment while in the respiratory tract of swine.[6] The last reported outbreak of avian influenza (H5N5) occurred in 1983–1984 in the eastern United States. The epidemic essentially eliminated the poultry population in the affected states. It should be noted that the disease did not eliminate the avian population entirely—this particular strain was killing 40–60% of the birds. The current methods to control and eradicate the disease were almost as devastating as the virus.[6]

From 1983 to 1994, the U.S. government has spent more than 60 million dollars combating the annual reappearance of variable pathogenic strains. Indirect losses to the industry were estimated to be 349 million dollars. The recent outbreaks in Hong Kong resulted in a conservative kill-off of 1.3 million birds. A 1998 outbreak in Mexico resulted in the eradication of 30% of breeder flocks and 50% of broiler flocks.

The influenza viruses of animals and humans can be considered very close cousins, having the capacity to infect both animals and humans. These viruses have evolved to exist in re-occurring infections in hosts ranging from young to adult to geriatric. They undergo a rapid rate of genetic mutation/variation resulting in both antigenic drift and antigenic shift. These terms imply that the genetic information (i.e., viral RNA) responsible for directing the synthesis of viral proteins is "sloppy" and often results in changes to some parts of the viral proteins. The constantly changing immune targets (variable epitopes) in the protein are preferentially recognized by the host immune system. This phenomenon does not allow the immune system to recognize other parts of the protein that are more conserved. Thus the immune system only really can effectively eliminate an identical strain of the virus it has been naturally infected with or vaccinated against. This viral survival mechanism complicates all of our current vaccination strategies. No current licensed avian influenza vaccine exists and experimental forms of the vaccines are not used at all in the United States. A procedure of monitoring for the virus and quarantine is the current control method.

The ease of infection, variable genetic makeup of the virus, and limited control methods illustrate why the U.S. poultry industry is at high risk for an intentional introduction of avian influenza. There should be a high research priority put on efforts to develop an effective vaccine against avian influenza viruses.

Velogenic Viscerotropic Newcastle Disease

Newcastle disease virus (NDV) belongs to the Paramyxoviridae virus family and, like other members of these group, possesses two surface proteins important to the identification and behavior of the virus.[1,7] The first, hemagglutinin/neuraminidase (HN), is important for the attachment and release of the virus from host cells. The second, the fusion (F) protein, has a critical role in pathogenesis of the disease. There are at least nine known types of avian paramyxoviruses based on the antigenic makeup of the hemagglutinin. NDV is the prototype virus for the Type 1 avian paramyxoviruses.

NDV occurs as three pathotypes: lentogenic, mesogenic, and velogenic, reflecting the increasing levels of virulence. The most virulent velogenic isolates are further subdivided into neurotropic and viscerotropic types. Velogenic viscerotropic Newcastle disease (VVDN) is also known as exotic Newcastle disease and Asiatic or Doyle's form of Newcastle disease.[1] VVDN is probably the most serious disease of poultry in the world. The lesions it produces in the gastrointestinal tract characterize the disease. In susceptible birds, disease rates approach 100% and mortality rates can exceed 95%. The 1971–1974 outbreak of exotic Newcastle disease in southern California provides a good example of the consequences of a killing disease in an intensive-farming situation. The disease was probably introduced by infected smuggled aviary birds. Once introduced into one of the most concentrated egg production areas in the nation, mortality approached 100%. It spread rapidly to commercial chicken and turkey farms. When initial control measures failed to contain the spread of disease, a state of national emergency was declared. Eradication of the disease involved the slaughter of nearly 12 million infected and exposed birds at a cost of 56 million dollars.

Within an infected flock VVND is transmitted by direct contact, contaminated feeding and watering equipment, and by aerosols produced by coughing, gasping, and other disturbances of respiration. Dissemination between flocks over long distances has resulted from the movement of contaminated equipment and service personnel, such as vaccination crews. Movement of asymptomatic infected birds accounts for most of the outbreaks in the pet bird industry. The most difficult part of eradication program is locating all infected and exposed birds. [It should be noted that it is not the policy of the USDA to recommend vaccination in the eradication of VVND.[1,7]]

Vaccines to other species-specific paramyxoviruses, Class I viral agents such as canine distemper, result in solid and durable protection. Vaccines against NDV using modified live vaccines are protective to varying degrees against lentigenic and mesogenic but not velogenic strains. Vaccination against the velogenic strains has been shown to protect a flock incompletely, resulting in the maintenance of infected carrier birds, which shed the virus at some other time. Also, vaccine strains in a flock may interfere with the interpretation of laboratory results and thereby delay the establishment of a diagnosis of VVND in the flock. Also complicating the picture is additional evidence for human spread of the disease. Although humans may become infected with Newcastle disease virus, the resulting disease is usually mild and transient. However humans may shed the virus 4 to 7 days following exposure and in some cases for more than 14 days.

BOVINE

Foot-and-Mouth Disease Virus (FMDV)

Foot-and-mouth disease (FMD) is an acute, highly contagious picornavirus infection of cloven-hoofed animals. This group of viruses constitutes a collection of animal and human viruses that exhibit both Class I agents (e.g., polio) and Class II agents (e.g., rhinoviruses, enteroviruses, and apthiviruses). FMD is present in many regions of the world, except for North and Central America (north of Panama), Australia, New Zealand, Great Britain, and Scandinavia. The European Union (EU) countries are generally free of FMD. FMD was last reported in the United States in 1929, in Canada in 1952, and in Mexico in 1954.[1]

The disease is highly contagious and may spread over great distances with the movement of infected or contaminated animals, products, objects, and people. Pigs are mainly infected by ingesting infected food. Waste feeding has been associated with outbreaks. The virus (FMDV) is sensitive to environmental influences, such as pH less than 5, sunlight, and desiccation. But it can survive for long period of time at freezing temperatures. Cattle are mainly infected by inhalation, often from pigs, which shed large amounts of virus in respiratory aerosols. Infected animals shed large amounts of virus before clinical signs are evident and winds may spread the virus over long distances.[1]

The incubation period of FMD is 2–21 days (3–8 days average), but virus is shed before clinical signs develop. The rate of infection (morbidity) can reach 100%, but mortality can range from 5% (adults) to 75% (suckling pigs and sheep). Recovered cattle may be carriers for 18–24 months and sheep for 1–2 months. Pigs are not carriers. In cattle, salivation, depression, anorexia, and lameness are the clinical signs. They are caused by the presence or painful vesicles (blisters) in the skin of the lips, tongue, gums, nostrils, coronary bands, interdigital spaces, and teats. In cattle, fever and decreased milk production usually precede the appearance of vesicles. The vesicles rupture, leaving large denuded areas, which may become infected secondarily. In pigs, sheep, and goats, the clinical signs are similar but milder. Lameness is the predominant sign.

Because of the range of species affected, the high rate of infectivity, and the fact that virus is shed before clinical signs occur, FMD is one of the most feared reportable diseases in North America. An outbreak of FMD would cost millions of dollars in lost production, loss of export markets, and loss of animals during eradication of the disease. Many other reportable diseases resemble FMD and it is important to distinguish between them at the earliest indications of an unusual disease outbreak. People can be infected through skin wounds or the oral mucosa by handling diseased stock, exposure to the virus in the laboratory, or by drinking infected milk. People cannot be infected by eating meat of infected animals. The human disease is temporary and mild. FMD is not considered a public health problem. But it remains the most economically significant animal disease with a major influence on the international trade of animals and their products.[1]

The world distribution of FMD has remained essentially unchanged in the last 20 years. This status quo has been maintained by balancing improved surveillance and diagnostic technology, the ever-increasing legal and illegal international movement of animals, and reductions in veterinary resources. FMD virus can now be cloned into a number of vectors and separately expressed and studied in isolation from the other viral proteins.[8] Biotechnology has not provided a safe and effective vaccine to replace the conventional tissue culture–derived vaccine. At any time, research on peptide, recombinant, and vector-expressed virus protein vaccines could yield a breakthrough, not only for FMD control but for control of other viral diseases, both human and animal. Until this occurs, control and eradication of FMD rely principally on classical epidemiological techniques with some new biotechnological methods.

The immune response to FMDV is type-restricted.[8] This means the immune system recognizes only one determinant on the virus. No two immune responses, induced via natural infection or vaccination, are similar enough to fully protect against the circulating "family" of viruses created by a high rate of viral mutation. The determinant for this dominant immune response is termed the G-H loop of FMDV within the VP1 protein. This is a highly mobile peptide that extends from the capsid surface. In native virions, it is invisible by X-ray crystallography. In serotype C, this segment contains a hypervariable region with several continuous, overlapping B-cell epitopes that embrace the conserved Arg-Gly-Asp (RGD) cell attachment motif.[8] A molecular model of an inserted stretch reveals highest flexibility of the RGD tripeptide segment compared with the flanking sequences, which could allow a proper accommodation to integrin receptors, even in poorly antigenic conformations. The non-converging structural requirement for RGD-mediated integrin binding and antibody recognition explains the dynamism of the generation of neutralization-resistant antigenic variants in the viral quasi-species. This might be of special relevance for foot-and-mouth disease virus evolution, since unlike in other picornaviruses, the cell binding motif and the major neutralizing B-cell epitopes overlap in a solvent-exposed peptide accessible to the host immune system in a virion lacking canyons and similar hiding structures. Currently, broadly protective immunity via vaccination is not feasible. A MASK technology, which has been developed for similar phenomena in HIV-1 infection, may now serve as a new avenue to develop an effective FMDV vaccine.[9]

Rinderpest

Rinderpest (RP), or cattle plague, is an acute, highly contagious viral disease primarily of cattle and buffalo and secondarily of sheep, goats, and wild ruminants.[1,10] The virus is in the genus Morbillivirus, which presently comprises measles virus of man, rinderpest virus (RPV), peste des petits ruminants virus (PPRV), and canine distemper virus (CDV). Rinderpest reached historic proportions one hundred years ago when a devastating pan-African epizootic started in the Horn of Africa in 1899. In only seven years, rinderpest spread down the length of the continent to South Africa. Approximately 80–95% of all cloven-footed animals (domestic cattle, sheep, goats, and pigs, along with buffaloes, giraffe, wildebeest, and antelope) were affected. In South Africa alone, 2.5 million cattle died before rinderpest was eradicated in 1905. Inflammation, hemorrhage, necrosis, and erosion of the digestive tract accompanied by wasting, and frequently bloody diarrhea characterized the disease. Mortality approached 90%. This pattern of pathogenesis is typical when fully virulent strains of the virus infect completely susceptible cattle populations. Pigs of European or North American origin may develop a mild infection with a mild transient fever, but they are capable of transmitting the virus to susceptible cattle.

Transmission is by contact with infected animals or indirectly with their secretions, excretions, and fomites.[1,10] The virus appears in the blood and secretions before clinical signs appear. Thus, the infection may be easily introduced inadvertently to slaughterhouses and stockyards. As is typical with this Class I viral pathogen, the few animals that recover develop solid immunity and a high antibody titer. They are not known to be carriers. Good vaccines exist to this group of viral agents, again, due to their genetic stability. An international program has been initiated to eradicate it from the world. In the United States, rinderpest is an FAD. The U.S. does have adequate reserves of vaccines for FMDV.

SWINE

African Swine Fever

African swine fever (ASF) is a highly contagious, generalized disease of pigs. It is caused by a large, lipoprotein-enveloped, icosahedral, double-stranded DNA virus (~200 nm) of the Family *Asfarvirida*. This new family name was required because the ASF virus straddles two distinct families of viruses. The morphology of the ASF virus is typical of iridoviruses, but the structure of the DNA (with terminal crosslinks and inverted terminal repeats) and the molecular biology of the virus are very similar to those of poxviruses.[1] ASF is currently the only member of this family and exhibits varying virulence between strains, although different serotypes cannot be identified. The ASF virus replicates in the cytoplasm of infected cells. It resists inactivation, and can persist up to 15 weeks in meat, 6 months in processed hams, and one month in contaminated pens. Domestic swine, wart hogs, giant forest hogs, and brush pigs are susceptible to ASF virus. Transmission and infection are the result of contact with infected, recovered, or carrier pigs; ingestion of contaminated garbage, urine, feces, and carcasses; or the bite of certain ticks

The virus probably existed in African wild boars, but disease was recognized by Montgomery in 1921, only after the introduction into Africa of European domestic pigs.[1,11] In 1957, the first outbreak of ASF outside of Africa was diagnosed in Portugal. Presumably pigs were fed infected swill from a meal served during a flight from Angola. This 1957 outbreak was apparently eradicated at the cost of 17,000 dead pigs, from the illness or from slaughtering because they had been in contact with sick animals. In 1960 the disease reappeared in Portugal and spread to Spain, becoming endozootically established in the Iberian Peninsula. ASF then appeared in the 1960s and early 1970s in France, Italy, and Cuba. In 1977–1978, the disease was detected in Malta, Sardinia, Brazil, Haiti, and the Dominican Republic. In most of these cases the disease was eradicated, with significant losses. It still exists in Sardinia. The latest appearances of the disease outside its endemic region were in Belgium in 1985 and in The Netherlands in 1986.

Initially, African swine fever always occurred as a fulminating, hyperacute disease with practically 100% lethality. After the disease became established in the Iberian Peninsula, acute and subacute forms of the disease became more frequent and are now the most commonly observed clinical forms of the disease. The clinical picture is not pathognomic and is easily confused postmortem with classical hog cholera (which in parts of Europe is referred to as swine fever or classical swine fever). Laboratory tests using direct fluorescent antibodies on suspected tissues are required to differentiate between the two diseases.

In Africa, warthogs and their associated ticks *(Ornithodoros moubata porcinus)* are the disease reservoirs for ASF, although ASF is not clinical in warthogs. Outbreaks occur when domestic pigs come in contact with the ticks, rather than from direct transmission from other pigs or warthogs. In Europe, direct transmission is more important as the virus is shed in high concentration in excretions and secretions during the acute phase. Ticks *(Ornithodoros erraticus)* on the Iberian Peninsula can also be a source of infection. Recovered pigs may remain chronically infected and excrete the virus for six weeks after infection. Contaminated pens and garbage feeding with material from international airports or seaports are documented methods of spread due to the resistance of the virus to inactivation. After an incubation period of 5–15 days, the disease may manifest itself in a number of forms: (1) peracute, pigs found moribund with death following rapidly; (2) acute, with high fever (up to 42°C), anorexia and recumbency, skin blotching, diarrhea, and abortion after 1–2 days (mortality is almost 100% within 7 days); (3) subacute, fluctuating or continuous fever for up to 20 days, with milder clinical signs; (4) chronic, transient recurring fever with stunting and emaciation, with possible pneumonia, lameness, skin lesions, and secondary infections.

An important, and unexplained, feature of ASF virus is the lack of induction of neutralizing antibodies. This has prevented the preparation of a conventional vaccine.

African swine fever is the most devastating disease of swine. It is a triple-threat to swine populations all over the world: (1) the virus does not induce the production of neutralizing antibodies and so conventional vaccine production has not been possible; (2) wild boars and ticks, which serve as disease reservoirs, are difficult to control; (3) healthy carriers or animals showing mild forms of the disease are difficult to diagnose and are appearing with increasing frequency. Thus, control of the spread of African swine fever still relies exclusively on rapid diagnosis, drastic slaughter, and effective sanitary barriers to the trade of pig or pork products.

From the viewpoint of vaccine development, the most striking aspect of ASF is the immune response's failure to induce neutralizing antibodies, both in pigs recovering from the disease and in laboratory animals inoculated with the virus.[1,12] This lack of production is not due to a generalized impairment of antibody production— animals do synthesize virus-specific antibodies. Yet in spite of the absence of neutralization, an immune protection seems to be elicited. Some surviving animals or animals inoculated with attenuated virus are resistant to reinfection with homologous virulent virus. It appears that the most virulent strains infect and rapidly destroy the antigen-presenting cells that would otherwise initiate the immune response. Without these cells the immune response cannot be effectively activated. Studies on immune functions of peripheral T lymphocytes revealed an abrogation of cellular immune responses as early as 3–5 days after infection and thus before detection of viral antigens in these cells. The data suggest that early immunosuppression represents a crucial event for the manifestation of classical swine fever. Cellular immunity may be the mechanism responsible for protection, since MHC-I–restricted cell-mediated cytotoxicity against infected macrophages has been demonstrated. In all forms of the disease, virus replication occurs mainly in mononuclear phagocytic cells. In lymphatic tissues macrophages and reticular cells are the main target cells, whereas in blood or bone marrow monocytes (and occasionally polymorphonuclear leukocytes and megakariocytes) are the main cells infected.

Hog Cholera

Hog cholera (HCV), a highly contagious viral disease of swine is caused by the lipid-enveloped virus of the family Togaviridae, genus Pestivirus.[1,13] Although minor antigenic variants have been reported, there is only one major serotype. HCV has a close antigenic relationship with the bovine viral diarrhea group and border disease virus. HCV is very stable in protein-rich environments and can survive for months in refrigerated meat and for years in frozen meat. It was eradicated from the United States in 1978 after a 16-year effort by industry and state and federal agencies. Today, only a few other countries in the world are free of hog cholera. In the spring and summer of 1997, outbreaks of hog cholera were confirmed in Haiti and the Dominican Republic, both of which had eradicated the disease in the early 1980s. Outbreaks also occurred in a number of countries of Europe with large losses reported in Belgium and The Netherlands. All these outbreaks had animal health officials at the U.S. Department of Agriculture (USDA) concerned that hog cholera could be introduced into U.S. swine herds. While hog cholera does not cause illness in people, economic losses to pork producers would be severe if the disease were to become established again in this country.

The most common method of transmission is direct contact between healthy swine and those infected with hog cholera. The disease can also be transmitted through contact with body secretions and excrement from infected animals. Healthy pigs coming into contact with contaminated vehicles, pens, feed, or clothing may contract the disease. Birds, flies, and humans can physically carry the virus from infected to healthy swine. Swine owners can inadvertently cause infection through feeding their herds untreated food wastes containing infected pork scraps.

The clinical signs of hog cholera vary with the severity of the infection. There are three forms of the disease: acute, chronic, and mild.

The acute form of hog cholera is highly virulent, causing persistent fevers that can increase body temperatures to as high as 107°F. Other signs of the acute form include severe depression, convulsions, and lack of appetite. Affected pigs will pile up or huddle together. Signs of hog cholera may not be apparent for several days after infection. Death usually occurs within 5 to 14 days following the onset of illness.

The chronic form of hog cholera causes similar clinical signs in affected swine, but the signs are less severe than in the acute form. Discoloration of the abdominal skin and red splotches around the ears and extremities often occur. Stunted pigs with chronic hog cholera can live for more than 100 days after the onset of infection.

The mild or clinically inapparent form of hog cholera seldom results in noticeable clinical signs. Affected pigs suffer short periods of illness often followed by periods of recovery. Eventually, a terminal relapse occurs. The mild strain may cause small litter size, stillbirths, and other reproductive failures. High mortality during weaning may also indicate the presence of this mild strain of hog cholera.

In countries where HCV is endozootic, a systematic vaccination program is effective in preventing losses.[1] Over the years, numerous regimens of vaccination have been advocated with a variable degree of success. In the past two decades, modified live vaccines with no residual virulence for pigs have become available. The Lapinized Chinese (C) strain, the Japanese guinea pig cell culture–adapted strain, and the French Thiveral strain have been widely used. All three strains are considered innocuous for pregnant sows and piglets older than two weeks. Should HCV be used against the United States in a terrorist act, no stores of vaccine are available to help control such an outbreak. Additional vaccine would have to be produced to provide enough doses.

SUMMARY

This paper was intended to highlight some of the disease agents that could be used effectively in acts of terrorism. In terms of vaccine countermeasures, we face situations on both ends of the spectrum—(1) we and other nations have not invested enough and have not been successful in developing or licensing any protective vaccines and (2) where vaccines are available but not commercially used due to current FAD policies we have not stockpiled them in sufficient doses should regular practices fail to contain an outbreak. It is hoped that this paper provokes additional thought and planning for those government agencies involved in the business of national food animal agricultural welfare. Vaccine technologies are available or are being developed to provide new and improved vaccines against these highly contagious agents.

REFERENCES

1. COMMITTEE ON FOREIGN ANIMAL DISEASES OF THE UNITED STATES ANIMAL HEALTH ASSOCIATION. 1992. Foreign Animal Diseases. U.S. Animal Health Association. Richmond, VA.
2. GORDON, J.C. & S. BECH-NIELSEN. 1986. Biological Terrorism: A direct threat to our livestock industry. Mil. Med. **151:** 357–363.
3. WATSON, W.A. 1986. Animal Health Today—Problems of large livestock units. Large livestock units and notifiable disease. Brit.Vet. J. **136**(1): 1–20.
4. Report to Congress on the Extent and Effects of Domestic and International Terrorism in Animal Enterprises. 1993. Physiologist **36**(6): 207–259.
5. Veterinary Biologics Notice No. 65. Biological Products in Licensed Establishments Produced and Destroyed 01/01/97 through 12/31/97. United States Department of Agriculture Animal and Plant Health Inspection Service. Biotechnology, Biologics and Environmental Protection. Hyattsville, MD. pp. 1–35.
6. WEBSTER, R.G. 1997. Influenza virus: transmission between species and relevance to emergence of the next human pandemic. Arch. Virol. Suppl. **13:** 105–113
7. U.S. DEPARTMENT OF AGRICULTURE. 1978. Eradication of Exotic Newcastle Disease in Southern California. 1971–74. Veterinary Services. U.S. Department of Agriculture.
8. DOEL, T.R. 1996. Natural and vaccine-induced immunity to foot and mouth disease: the prospects for improved vaccines. Rev. Sci. Tech. Epiz. **15**(3): 883–911.
9. GARRITY, R.R., G. RIMMELZWAAN, A. MINASSIAN, W.P. TSAI, G. LIN, J.J. DE JONG, J. GOUDSMIT & P.L. NARA. 1997 Refocusing neutralizing antibody response by targeted dampening of an immunodominant epitope. J. Immunol. **159**(1): 279–289.
10. BARETT, T. 1996. Morbilliviruses in the twenty-first century. *In* Proceedings of the FAO Technical Consultation on the Global Rinderpest Eradication Programme held in Rome. pp. 25–38.
11. DAHLE, J. & B. LIESS. 1992. A review on classical swine fever infections in pigs: epizootiology, clinical disease and pathology. Comp. Immunol. Microbiol. Infect. Dis. **15**(3): 203–211.
12. MARTINS, C.L. & A.C. LEITAO. 1994. Porcine immune responses to African swine fever virus (ASFV) infection. Vet. Immunol. Immunopathol. **43**(1–3): 99–106.
13. MOENNIG, V. 1992. The hog cholera virus. Comp. Immunol. Microbiol. Infect. Dis. **15**(3): 189–201.

Exotic Diseases of Citrus
Threats for Introduction?[a]

RONALD H. BRLANSKY[b]

University of Florida, Citrus Research and Education Center, Lake Alfred, Florida 33850, USA

THE U.S. CITRUS INDUSTRY

The citrus industry in the United States is an important agricultural industry and is primarily located in the states of Florida, California, Texas, and Arizona. The total acreage for each state is shown in TABLE 1. This represents a total U.S. acreage of more than 1.20 million acres. The on-tree value for all citrus in these areas in 1996–1997 was more than $1.3 billion. This figure accounts for the total value of the crop after other factors are figured.

TABLE 1. United States Citrus Production Areas by State in 1996–1997

Arizona	California	Florida	Texas
34,800 acres	272,800 acres	857,861 acres	32,800 acres

In the past, moving citrus planting materials from area to area within the United States and from outside the United States was routine. This routine movement introduced foreign diseases and pests into new areas. Even though this movement is now severely restricted, some new diseases and pests have appeared. The purpose of this paper is to discuss the foreign diseases of citrus that pose a threat to citrus areas of the United States and the appropriate cooperative research needed to support means of early detection and spread.

TYPES OF CITRUS DISEASES

Citrus diseases are not different from the diseases of other plant species. The types of pathogens that cause these diseases are fungi, bacteria, viruses, phytoplasmas, spiroplasmas, and nematodes. Citrus disease symptoms can be leaf and fruit spots, cankers, rots, leaf chlorosis, dieback, stunting, small fruit, misshapen fruit, and mottling of leaves resembling nutrient deficiencies.

[a]Florida Agriculture Experiment Series No. R-05780.
[b]Address correspondence to: Ronald H. Brlansky, University of Florida, Citrus Research and Education Center, 700 Experiment Station Road, Lake Alfred, Florida 33850; Telephone: 941-956-1151; Fax: 941-956-4631.
e-mail: rhby@tangelo.lal.ufl.edu

PATHOGEN TRANSMISSION

Each pathogenic agent spreads in a different manner. Fungal or bacterial citrus pathogens are transmitted by propagules that can infect citrus plants through natural openings or wounds. Fungi are spread by mycelia, spores, and other fruiting bodies. Bacteria are spread only by bacterial cells. Many systemic pathogens of citrus (i.e., viruses, phytoplasmas, spiroplasmas, and bacteria) can be contained in propagation materials (buds or stocks) of the plant. Some may not produce symptoms on the initial host plant but cause disease when introduced into the appropriate citrus host or cultivar. Most of these systemic pathogens are also transmitted by specific insect vectors that feed on citrus plants. Information is often lacking on the ability of the pathogens to survive in the insects and the efficiency of the insects to successfully transmit the pathogens. Factors such as temperature, humidity, and plant host affect transmission ability.

FOREIGN OR EXOTIC CITRUS DISEASES OF CONCERN

A list of six citrus diseases and the causal agent of each is shown in TABLE 2. These diseases do not exist in the continental United States yet. Two of these diseases, citrus variegated chlorosis and citrus greening, are discussed in detail because both are present in major citrus growing areas of the world and both cause serious problems in citrus production.

Citrus variegated chlorosis was first reported in Brazil in 1987. The disease is caused by the bacterium *Xylella fastidiosa*, which is present only in the xylem of infected plants. The disease is present in all major citrus areas of Brazil. It causes an interveinal chlorosis of leaves along with brown lesions in the centers. Infected fruit ripen early and are undersized and hard. It has been reported that in Sao Paulo state 34% of 150 million citrus trees have symptoms of the disease. Sharpshooter leafhoppers (Cicadellidae), which feed on the xylem, are vectors of the pathogen. A sharpshooter leafhopper present in Florida has been shown to vector the bacterium to citrus. This disease poses a serious threat of being introduced into Florida and it could cause considerable damage.

Citrus greening is a serious disease of citrus in many areas of Asia and Africa. The disease is caused by a systemic phloem-limited bacterium. The disease has been called many names such as likubin, vein-phloem degeneration, leaf mottling, and huanglongbing. Two forms of the disease exist: the Asian form (most active in warm conditions of up to 32°C) and the African form (most active in cool conditions of 20-24°C). Both are vectored by psyllids. The Asian form is vectored by the citrus psyllid, *Diaphorina citri*, which has recently been found in Florida. Symptoms of the disease include yellowing of shoots, leaf vein chlorosis, and leaf mottling with an eventual dieback of limbs. Fruits on infected trees are small and poorly colored. The juice is low in soluble solids, high in acid and therefore bitter. This disease can cause severe symptoms is young trees before they can bear fruit. In some citrus areas where greening exists citrus production has become difficult.

Exact figures on yield reduction due to many of these diseases are unavailable. Control of a systemic disease is difficult since many citrus types do not exhibit

TABLE 2. Exotic citrus diseases of importance

Disease	Causal Agent	Vector/U.S. Presence	Detection Means
Greening	Phloem bacterium	Psyllid/Yes	Hybridization
Citrus variegated chlorosis	*Xylella fastidiosa* (Xylem bacterium)	Leafhopper/Yes	Culture, serology, PCR
Citrus tristeza virus (Stem pitting form)	Closterovirus	Aphid/Yes	Host index, serology, PCR
Black spot	*Guignardia citicarpia*	None	Symptoms, culture
Citrus chlorotic dwarf	Virus?	Whitefly/Yes	None
Witches broom of lime	Phytoplasma	Leafhopper/No	Hybridization

symptoms and yet are sources of the pathogen for insect vectors to spread to susceptible cultivars.

DISEASE/PATHOGEN DETECTION

Early methods for pathogen detection have been based on host reactions. Detection of these diseases is normally based on detection of the pathogen since symptoms can be confused with other factors often due to non-pathogenic causes. Pathogen detection for fungal and bacterial pathogens has been based on culturing and inoculation of the causal agent. Serology (antibody based) is used to overcome (1) the difficulty of culturing fastidious (slow growing, nutrient specific) prokaryotic pathogens, such as *Xylella fastidiosa,* and (2) the inability to culture the greening bacterium and viruses, such as citrus tristeza virus. Nucleic acid–based methods (hybridization and polymerase chain reaction) have also been devised, but are expensive, require specialized equipment, and are not yet widely available for diagnostics.

INTERNATIONAL COOPERATIVE RESEARCH

There continues to be considerable international cooperative research on many of these exotic citrus diseases. Scientists worldwide see the benefits from such cooperative research. Scientists from countries where these diseases exist have the opportunity of learning new methodologies that can be applied to disease control, while scientists from those areas where the disease is absent can develop methods useful for early detection. Work with insect vectors may lead to more understanding of these diseases. Much of this cooperative research is funded by agencies such as the U.S. Department of Agriculture's Foreign Agricultural Service and Research and Scientific Exchanges and the National Science Foundation's International Cooperation Grants.

PORTS OF ENTRY AND DEFENSES

With the increased frequency of air travel between foreign countries and the United States, the frequency of interceptions of foreign plant materials and insects has also increased. Take, for example, the most recent introduction of the citrus psyllid *Diaphorina citri* (Kuwauama) into Florida—since 1985 the psyllid was intercepted 40 times at U.S. ports of entry! Other species of the same genus were also intercepted on rutaceous plants from Asia. The brown citrus aphid, *Toxoptera citricida,* (Kirkaldy), the most efficient aphid vector of severe citrus tristeza virus, was also newly discovered in Florida. Continual cooperative efforts are warranted among government, university researchers, and inspection agencies.

What Should the G8 Do about the Biological Warfare Threat to International Food Safety?

WENDY BARNABY
Journalist, Brandreth, Station Road, Chilbolton, Stockbridge, Hants S020 6AW, U.K.

BIOLOGICAL WARFARE: WHAT, WHO, AND WHY

Biological warfare (BW) is the deliberate use of living microorganisms to cause disease or death in people, animals, and plants. Two countries are known to have biological weapons capabilities: Russia and Iraq. Both the U.K. and the U.S. carried out research into offensive biological capabilities and weaponized agents before they renounced their offensive programs—the U.K. in 1957, the U.S. in 1969. However, Iran, Israel, Libya, Syria, China, North Korea, and Taiwan are generally assumed to have a biological capability. South Africa has recently admitted its research into BW.

The main attraction of a BW capability is its relative cost-effectiveness. Evidence presented to a United Nations panel in 1969 estimated that "for a large-scale operation against a civilian population, casualties might cost about $2000 per square kilometer with conventional weapons, $800 with nuclear weapons, $600 with nerve-gas weapons and just $1 with biological weapons."[1] No wonder biological weapons have been called "the poor man's nuclear bomb."

Not surprisingly, most writings about BW concern the targeting of people. Yet there has also been a great deal of research into targeting crops. The attraction of an anti-crop capability is the tremendous damage it could inflict on the enemy.

How Damaging?

The Irish potato famine of 1845–1846 was a brutal example of the disasters that befall a population when staple crop fails. Late blight of potatoes brought starvation and emigration in its wake. The natural mechanics of the spread of crop diseases mean that, once released, disease-causing organisms are disseminated quickly and widely. The fact that outbreaks of crop pests and diseases happen naturally makes it harder for an enemy to pin down crop losses as the result of attack rather than of natural disaster. BW targeting crops or food could be carried out by governments or non-government groups such as terrorists.

GOVERNMENT-SPONSORED BIOLOGICAL WARFARE

Governments are more likely to attack the crops of an adversary rather than their produced food. Targeting crops would produce more disruption more easily and probably with no definite blame attached to the aggressor. Most is known about the anti-crop BW programs carried out by the United States and Iraq.

The United States

The U.S. Chemical Warfare Service wrote periodic appraisals of "bacteriological" warfare, but planned development was not recommended until 1942.[2] Strains of pathogens were selected, their nutritional requirements evaluated, optimal growth conditions and harvesting techniques developed, and a form suitable for dissemination prepared. Extensive field testing assessed their effectiveness on crops.[3]

By the time President Nixon renounced the U.S. offensive biological research program in 1969, the country had standardized five anti-crop BW agents. The identities of two of these agents have not been revealed. The three thought to have been produced in the greatest quantities were (1) stem rust of wheat (*Puccinia graminis tritici*), which formed a stockpile of 36,000 kg; (2) *Tilletia indica,* which causes extensive damage to rice in its later stages of development and to barley, corn, rye, and wheat as seedlings, and (3) a stockpile of 900 kg of rice blast pathogen (*Piricularia oryzae*).[4] Research into delivery systems considered, according to the U.S. Department of the Army, "burster type bombs..., submunitions, gas explusion [*sic*] bombs, various types of line source spray tanks and highly specialized projectiles and generators, as well as insect vectors."[5] One of the dispersion systems was bizarre indeed: feathers dusted with pathogens and released from aircraft, which were shown in tests to carry enough spores to cause a cereal rust epidemic, and which were also designed to disseminate anti-crop pathogens from unmanned free-floating balloons. Spray tanks fitted to aircraft were another means of dispersal to be tested. It has been argued that a single aircraft filled with spray tanks could initiate plant disease epidemics over areas "greatly in excess of 1,000 km squared from a downwind distance of some 30 miles."[6]

Iraq

The extent of Iraq's biological weapons program uncovered by United Nations inspectors following the Gulf War in 1991 took everyone by surprise. As well as the bombs and warheads filled with botulinum, anthrax, and aflatoxin, Iraq developed a fungal pathogen called *Tilletia indica*, one of the agents also developed by the U.S. offensive program. It causes a range of diseases known as rusts, blasts, and smuts in cereals such as rice, wheat, oats, and corn.[7] Iraq may have been particularly interested in attacking wheat, the main cereal crop in Iran. According to Whitby and Rogers, "The fungus attacks the flower of the wheat plant and replaces the developing seed with masses of black teliospores. As well as substantially decreasing yields of wheat crops, it produces trimethylamine gas which can cause explosions in harvesters."[8]

In each of these national programs, anti-crop capabilities were incorporated into the general effort. They could certainly have inflicted enormous economic damage on an enemy.

Future National Programs

As mentioned above, there are at least seven countries assumed to be conducting BW research. As far as scientific expertise goes, they are middling countries with some scientific infrastructure, but not at the forefront of research. The danger with BW proliferation is that this level of sophistication is entirely adequate, as Iraq has proved. Weaponizing an agent is more difficult than growing it; but even crude weapons can be effective.

There is no doubt that any middling state that wanted to could produce anti-crop biological weapons. The only factor restraining their use would be the physical proximity of enemy crops. Anti-crop BW would not be efficient in internal wars between governments and insurgents, as there would be a great danger of contaminating one's own crops as well as the enemy's. However, where there is some geographical separation, the method would be cheap and effective.

NON-GOVERNMENTAL BIOLOGICAL WARFARE

While national governments would see the utility of poisoning food crops, contaminating food itself would not produce enough casualties quickly enough to be a form of national warfare. It is far more likely to be used by non-governmental groups such as terrorists, who generally want to create newsworthy public fear and disruption. Food poisoning is dramatic and visible, drawing public attention to the group and its demands. There have already been two well-documented incidents of food contamination with biological agents.

In September/October 1984, members of the Baghwan Shree Rajneesh cult contaminated salads in outlets in The Dalles, Oregon, with *Salmonella typhi*, which causes typhoid fever. Nobody was killed as a result of the contamination, but 750 people became ill. [9]

The other incident occurred in October/November 1996, at a large medical center in Texas. Laboratory staff who accepted an invitation to eat cakes left in their break room came down with nausea and diarrhea, later diagnosed as dysentery caused by the laboratory's own culture of *Shigella dysenteriae*. [10]

WHAT ADVICE COULD WE GIVE TO THE G8?

Action against terrorists

Given all of these developments, what can the G8 governments do to minimize the risks of BW for their populations? All countries should strengthen their domestic legislation on the handling of dangerous pathogens. The United States has realized the need to safeguard stocks of cultures. Regulations (Antiterrorism and Effective Death Penalty Act of 1996, signed into law on April 24, 1996) which went into effect in April, 1997, put an end to the casual supply of agent from culture collections. Individuals such as the Aryan supremacist Larry Harris can no longer order bubonic plague bacteria from the American Type Culture Collection in Maryland and have it shipped to him by Federal Express. The new regulations, drawn up by the Centers for Disease Control and Prevention in Atlanta, restrict transfers of such agents and the institutions that may handle them.

The U.K. has passed the 1974 Biological Weapons Act to prohibit possession of suspect biological agents or development of biological weapons by any person or group. The penalty for these offenses is life imprisonment. So far, there have not been any prosecutions under the act.

Action against Use by Nations

The Biological Weapons Convention (BWC) is the current focus of nations' efforts to control the spread of biological weapons. It was opened for signature in 1972 and entered into force in 1975. At the moment, 158 countries have signed and, of these, 140 have ratified it. The convention is the first to outlaw a whole class of weapons. Yet it has no provisions at all for checking that the parties are keeping to their obligations. Given the fact that the former Soviet Union admitted in 1992 that it had maintained its offensive biological weapons program in spite of its ratification of the BWC, and the suspicion that at least seven other nations that have also ratified may well be developing their own BW capabilities, the lack of any verification system is crippling to the convention's effectiveness. It is a problem the members of the Convention are grappling with: they have appointed an *ad hoc* group (AHG) to come up with a legally binding, workable regime.

The AHG is working on a Verification Protocol that will have four main provisions. Parties would make mandatory declarations about the work being done in their biotechnological establishments: biodefense facilities, vaccine producers, maximum containment labs, etc. There would be a system of (hopefully) non-confrontational inspection of these facilities to ensure, as far as possible, the accuracy of the declarations. If there were some worry that a party was not complying with its obligations, challenge inspections would come into play. There would also be investigations of allegations of use of biological weapons.

The AHG is working with lists of pathogens that parties would be required to declare. As far as anti-crop agents go, the list includes 20 that attack grains, coffee, potatoes and other root vegetables, citrus trees, and economic targets, such as pine trees. The list also includes two insects. One insect, *Thrips palmi*, was added in response to Cuban accusations that the U.S. sprayed the insect onto Cuban potato crops in 1997. (The U.S. denies the charge.) Genetically modified organisms are also cited. This list, like the others the AHG is working with, is meant to be illustrative rather than comprehensive.

One of the problems that has dogged the conclusion of a verification protocol is the proposed system of inspections. The pharmaceutical industry in both the U.S. and the U.K. has felt very threatened by on-site inspections, worried that they might compromise commercial secrets. These worries have been addressed in the U.K. by pharmaceutical plants volunteering for mock inspections. These caused problems, foreseen and unforeseen; but they led to an understanding between industry and government of each other's positions and problems, and a willingness to work together for solutions.[11] The attitude of U.S. industrialists has been even more strongly opposed to the idea of inspections and needs to change if an effective verification protocol is to be negotiated. At the current rate of progress a protocol may be ready by 2001, which will also be the time of the next conference in the series called periodically to review the effectiveness of the BWC.

Biodiversity

An effective verification regime is necessary not only because of the dangers of proliferation, but also because of the dangers of decreasing biodiversity. It has been estimated that we are now losing a minimum of about 50,000 species a year from a planetary complement of 10 million species. This extinction rate is accelerating so rapidly that by the middle of the next century one-half of all species and most populations of surviving species may be gone.[12] Food crops are constantly improved by germ plasm from wild relatives of modern crops, land races, and so-called primitive cultivars. If these should be threatened, the productivity increases (of about 1% annually) they account for will also be sacrificed.[13] Both wheat and rice have become genetically impoverished with the drive towards genetic uniformity.

A country that is almost entirely dependent upon a single grain crop and possibly a single cultivar of the single grain is very vulnerable to attack by a pathogen able to overcome resistance in that crop. In these days of genetic engineering of pathogens, such scenarios become more frightening as a method of biological warfare. Mixed cropping is less dangerous and there should be more of it. Quarantine procedures have a role to play here too, to minimize the spread of plant disease. Integrated pest management, involving pesticides, biological control agents, and resistant cultivars, should also be strengthened. These are measures the G8 can put in place.

Responsibility of Scientists

Biological weapons are a perversion of biology, and all professional biologists should be concerned to keep their subject from plumbing such depths. While some are actively involved, it is a vanishingly small minority. Few working biologists are interested—an impression given substance by a poll of ten biological societies in the U.K.[14] The G8 would do well to encourage scientific societies to follow the example of the Royal Society, which is the only one to have addressed any BW issues. In 1994, the Royal Society produced a report "to examine how science could contribute to the control of biological weapons." Speaking personally, its then President, Sir Michael Atiyah, said, "I believe that scientists should speak out on matters such as these. It would be immoral not to."[15]

ACKNOWLEDGMENTS

I would like to thank Simon Whitby of the Department of Peace Studies, Bradford University, U.K., and Dr. Annabelle Duncan of CSIRO, South Clayton, Victoria 3169, Australia, for their help in the preparation of this paper.

REFERENCES

1. Chemical and Bacteriological Weapons and the Effects of their Possible Use. 1969. United Nations. New York.
2. WHITBY, S. & P. ROGERS. 1997. Anti-crop Biological Warfare—Implications of the Iraqi and US programs. Defense Analysis **13**(3): 303–317.
3. LAUGHLIN, JR., L. 1977. U.S. Army Activity in the U.S. Biological Warfare Programs. Vol. II. Department of the Army. C2.
4. ROBINSON, J.P.P. 1981. Environmental effects of chemical and biological warfare. *In* War and Environment. W. Barnaby, Ed. Environment Study Council. Stockholm.
5. LAUGHLIN, op. cit.
6. ROBINSON, op. cit.
7. WHITBY & ROGERS, op. cit. p. 305.

8. Ibid.
9. TOROK, T.J. *et al.* 1997. A large community outbreak of Salmonellosis caused by intentional contamination of restaurant salad bars. J. Amer. Med. Assoc. **278:** 389–395.
10. KOLAVIC, S.A. *et al.* 1997. An outbreak of *Shigella dysenteriae* Type 2 among laboratory workers due to intentional food contamination. J. Amer. Med. Assoc. **278:** 396–398.
11. BARNABY, W.E. 1997. The Plague Makers: The secret world of biological weapons. pp. 157–174.Vision Paperbacks. London.
12. MYERS, N. 1998. Population dynamics and food security. *In* Food Security: New solutions for the 21st century. S.R. Johnson, Ed. Iowa State University Press. Ames, Iowa.
13. Ibid.
14. BARNABY, op.cit. pp.188–192.
15. ATIYAH, M. 1995. Anniversary Address. Royal Society. London.

Roundtable Summary

A Domestic Legislative Agenda for Improving Food Safety and Safeguards from Terrorist Attacks on the U.S. Food Supplies and U.S. Agricultural Interests

RAPPORTEUR: LONNIE KING[a]

Michigan State College of Veterinary Medicine, East Lansing, Michigan 48824-1314, USA

Humanity has always lived with, often barely surviving, plagues, pestilence, and epidemics. Today, horrors such as the Black Death, which wiped out almost half the European population in the fourteenth century, are distant historical events. Vaccines, antibiotics, and extraordinary advances in biomedical science and technology in this century have created the impression that we are all but impervious to the ravages of disease.[1]

Yet, diseases once under control, like tuberculosis and malaria, are beginning to resist antibiotics and new, emerging pathogens are now found in our hospital wards. The impact of AIDS on an entire generation and the potential effects of approximately 30 new or reemergent pathogens have made us reconsider the status of our public health. The socioeconomic, demographic, and political changes in the last 20 years have created a new milieu favoring, once again, the adaptiveness and persistence of the microbial world.

In addition, since the 1970s, the world has seen a phenomenal growth in terrorist organizations and the increased presence of a dangerous minority of national leaders with huge ambition and little scruples. Biological weaponry has become disturbingly attractive to both.

Since the revolutions in biotechnology in the 1970s and 1980s made it possible to breed and manipulate microbes as never before, the skills, knowledge, and equipment required to produce biological weapons may depend only on the competence of a biology graduate, in contrast to the substantial resources required to produce nuclear weaponry.

Whether we like it or not, modern warfare is no longer limited to the battlefield. The goal of conflicts reach beyond the combat zone and include the disruption or destruction of an enemy's supporting social structure. Biological weapons (BW), at least in theory, give military commanders the ability both to put soldiers out of action on the battlefield and cripple civilian populations, either directly or indirectly, through targeting their wealth, industry, and infrastructure, which sustain their military resources.

The U.S. has probably the most developed and sophisticated agriculture system in the world. The disruption of our food system could devastate not only our econo-

---

[a]Address correspondence to: Dr. Lonnie King, G100 Veterinary Medical Center, Michigan State College of Veterinary Medicine, East Lansing, Michigan 48824-1314; Telephone: 517-355-6509; Fax: 517-432-1037.

my but also deprive us and our military of basic resources, such as food, and bring chaos to essential services. A handful of strategically placed biological agents or toxins can spread infections and disease among livestock or ruin huge areas of crops in a way that could not be duplicated by conventional weapons. The use of BW targeting at our key crops or livestock, which would help undermine our economy, has the added advantage of possibly being misconstrued as a natural misfortune rather than a clandestine act and being overlooked completely.

The focus of this roundtable has been on setting a domestic legislative agenda for improving food safety and safeguards from terrorist attacks on our food supply and our agricultural interests. Perhaps we should address this topic in more practical terms, i.e., what are the compelling reasons that Congress and state legislatures should put bioterrorism on their radar screens and what recommendations should be coordinated into a new plan of action to address this pressing issue?

As with most complex and national issues, bioterrorism and BW need to be viewed from a number of perspectives and when put together, produce a new, kaleidoscopic picture that is much different from that produced by looking through a single lens. The views of this panel come from academia, veterinary medicine organizations, the food animal industry, as well as a variety of military and federal government points of view.

Dr. Boyle, executive vice president of the National Association of Federal Veterinarians, reviewed the type of agents that might be favored in BW, the characteristics of those agents that favor their use, and the potential public reaction and concern. He further suggested that the extremely mobile population of the U.S. makes us more vulnerable to BW attacks. He recommended improvements in preparedness, diagnostic capabilities, disease control, security, prevention, and in our epidemiological skills.

Dr. Finnegan, director of the Government Relations Division of the American Veterinary Medical Association, described what is at risk both in human and food animal populations. He discussed the trends in agricultural production including "factory farming," i.e., the concentrations of large numbers of food animals in the hands of fewer and fewer producers, and the phenomenal mobility and movement of animals and their products both nationally and internationally in a relatively short period of time. These trends give us a new respect for the concept of exposure potential. Dr. Finnegan also suggested that both our public health and veterinary infrastructure is not "tuned into" seeing or diagnosing exotic or seldom seen pathogens. Together these factors could lead to a scenario where a BW with a relatively long incubation period could expose large populations, often with little immunological protection, over large geographical settings. Diagnosis would rely on a relatively inexperienced group of health professionals.

Joel Brandenberger, vice president for legislative affairs at the National Turkey Federation (NFT), pointed out that our response to bioterrorism suffers from the same inadequacies currently facing our national food safety system: the system is fragmented and nonintegrated, based on antiquated federal statutes. As an industry representative, Mr. Brandenberger expressed concern about the potential risk from animal-rights extremists and the need for clearer and stronger laws to both prevent and punish terrorists threatening our food supply.

Dr. Sandy Miller, Adjunct Fellow at Georgetown University's Center for Food and Nutrition Policy, reviewed the technological sophistication of the food industry and, at the same time, the vulnerability of the industry to contamination of products at the processing, storage, and distribution points along the food chain. He further emphasized the need for risk-based food safety systems and the need for greater knowledge and application of risk analysis methodologies. Dr. Miller pointed out the need for stronger partnerships and linkages among federal, state, industry, and university resources to upgrade our research, education, and responsiveness to BW threats.

American agriculture is a phenomenal industry and our affordable, abundant, nutritious, and relatively safe food supply is the envy of the world. Americans only spend 11% of their disposable income on food—a bargain worldwide. In addition, such outstanding food systems play an important role in the foundation of our national economy and the social stability of our population. Post GATT and WTO, agricultural trade has enjoyed a tremendous growth. The U.S. exports almost $60 billion of agricultural products annually and imports almost half this amount for our own consumers. Almost one of every five jobs in the U.S. jobs is associated with agriculture and our expanded food system. It is easy to ascertain the substantial impact of any threat or disruption to this system.

With so much at stake and so much at risk, this panel is deeply concerned and laments our current condition that leaves us inadequately prepared to prevent, detect, deter, and respond to a bioterrorist or BW attack.

A decade ago, the National Academy of Sciences published the book, *The Future of Public Health*. The text emphasized the fact that the public has come to take the success of public health for granted and that this complacency has lead to decreases in resources and to a public health infrastructure in disarray. Ten years later, this same infrastructure may be even less prepared, in greater disarray, yet faced with far greater risks.[2]

In 1994, the Centers for Disease Control and Prevention (CDC) published a strategic plan to address emerging infectious disease threats. The plan cited the need for improved surveillance, better applied research, and more effective prevention activities to maintain a strong defense against disease agents threatening our public health.[3] While some progress has been made, we have fallen woefully short of supporting the goals of CDC's action plan.

Dr. D.A. Henderson, distinguished Johns Hopkins professor and former chief of CDC's Surveillance Section stated, "We need to be as prepared to detect and diagnose, to characterize epidemiologically and to respond appropriately to biological weapons use as we need to be prepared to respond to the threat of new and emerging infections." Henderson warned "Knowing what little has been done to date, I can only say that a mammoth task lies before us. If we can and are willing to spend tens of billions to deal with the threat of nuclear weapons, as now is the case, we should be more than prepared to devote hundreds of millions to cope with the greater threat of new and emergent infections whether naturally occurring or induced by man."

BW can target human, animal, and plant sources with equally devastating results. The public's perceptions, emotionalism, and penchant for sensationalism in the news turn small threats or actions into instantaneous national events. We only need to follow the bovine spongiform encephalopathy (BSE) stories in Europe or our responses to the Hudson Foods and Jack-in-the Box food safety incidents and their aftermaths to understand what an anthrax, plague, or botulism incident might do to the public psyche. It seems as if past warnings have not been heeded. The first step in setting a legislative agenda is to get the attention of decision-makers so that they will be receptive to any initiative.

In a recent book published by the National Academy of Sciences, *Ensuring Safe Food from Production to Consumption*,[4] the authors defined the attributes of an effective food safety system. These attributes not only help lay the groundwork for building an effective food safety system but they also can be used as parameters for a national system to protect and respond to BW. An effective system should be: (1) science based with a strong emphasis on risk analysis and prevention; (2) based on clear, rational national law; (3) supported by a comprehensive surveillance and monitoring system (4) headed by a single, central focal point at the federal level that is responsible for food safety (or bioterrorism) and has the authority and resources to implement science-based policies and programs; (5) supportive of the roles and responsibilities of non-federal partners (state, local, industry, etc.); and (6) adequately funded to carry out essential functions and missions. Our roundtable panel emphasized these points. These six attributes should form the basis for any legislative agenda. Perhaps the idea of being able to simultaneously improve both our food safety system and biological warfare response would be an attractive approach in establishing our legislative agenda.

In addition, a legislative agenda must support new efforts to reconnect our rapidly changing and global food system with a modern food safety system put together to address new challenges such as BW. Statutes and organizational changes must also be addressed immediately. It is essential to create rapid-response teams and to train and hire more professionals with strong clinical, diagnostic, microbiological, and epidemiological skills. The public health and animal health infrastructures must be rebuilt in order to deal with BW and emerging disease agents. Not only will this improve our capacity to counter the genuine threat of bioterrorism, it, simultaneously, helps respond to our new vulnerabilities to naturally occurring infectious diseases.

There needs to be emergency planning that is well-coordinated among federal, state, and local agencies. Stronger surveillance systems with technologically advanced biosensors and other improved diagnostic tools, improved training and research, and efforts to streamline systems are essential to effectively pre-empt, deter, and respond to new threats.

According to the World Tourism Organization, 528 million people vacationed in foreign countries in 1994. Americans accounted for 47 million of this total and more than 43 million tourists came into the U.S. in 1995 alone. Business travelers add substantially to this total. Armed with this knowledge and a better scientific understanding of BW, we now have a new appreciation of what a contemporary "Trojan horse" might be.

Enemy lines can be penetrated by new types of BW and with new or reemergent payloads capable of exposing pristine and vulnerable human, animal, and plant sources. Only an aggressive, science-based and comprehensive legislative agenda is acceptable. Laws, regulations, capacities, and organizations will need to be reinvented, strengthened, and refocused on contemporary threats and issues. Anything else is unacceptable in the face of these new threats and their substantial consequences.

REFERENCES

1. BROOKESMITH, P. 1997. Biohazard—The Hot Zone and Beyond. Barnes and Noble, Inc. New York.

2. INSTITUTE OF MEDICINE. 1988. The Future of Public Health. National Academy Press. Washington, D.C.
3. CENTERS FOR DISEASE CONTROL AND PREVENTION. 1994. Addressing Emerging Infectious Disease Threats—A Prevention Strategy for the United States. National Center for Infectious Diseases. Washington, D.C.
4. INSTITUTE OF MEDICINE. 1998. Ensuring Safe Food from Production to Consumption. National Academy Press. Washington, D.C.

Index of Contributors

Adams, J.B., 73–75
Alibek, K., 18–19

Bandyopadhyay, R., 28–36
Barnaby, W., 222–227
Bickerton, G., 44–47
Breeze, R.G., 9–17
Brlansky, R.H., 218–221
Brown, B., 92–94
Buntain, B., 44–47

Courtney, B.C., 130–148

Deen, W.A., 164–167
Douglass, J.D., Jr., 118–123
Dunn, M.V., 184–188

Ezzell, J.W., 130–148

Fitzgerald, R.B., Jr., 149–153
Forsythe, K.W., Jr., 95–99
Franz, D.R., 100–104
Frazier, T.W., 1–8
Frederiksen, R.A., 28–36

Garrity, R.R., 124–129

Henchal, E.A., 130–148
Hickson, R.D., 108–117
Higgins, J.A., 130–148
Horn, F.P., 9–17
Huxsoll, D.L., 105–107

Ibrahim, M.S., 130–148

Kijek, T.M., 130–148
Kildew, J.A., 37–43
King, L., 228–232

Knauert, F.K., 130–148

Lautner, B., 76–79
Loomis, L., 168–180
Ludwig, G.V., 130–148

Madell, M.L., 95–99
Murphy, F.A., 20–27
Myers, M., 168–180

Nara, P.L., 206–217
Neher, N.J., 181–183
Nelson, A.M., 83–91
Noah, D.L., 37–43

Olson, K.E., 68–72
Ostroff, S.M., 37–43

Probst, P.S., 154–158

Ragsdale, N.N., 199–205

Seitzinger, A.H., 95–99
Sequeira, R., 48–67
Sobel, A.L., 37–43
Stolte, D., 68–72

Thompson, B.H., 189–198
Torres, A., 80–82

Valdes, J.J., 168–180
von Bredow, J., 168–180

Wagner, D., 168–180
Watson, S.A., 159–163

Zamani, K., 168–180